The Story of
FIREFIGHTING
in Canada

The Story of FIREFIGHTING *in Canada*

by

Donal M. Baird

The spectacular fire in Ordnance Square, Halifax, was one of the many that were fought by a combination of town and military fire engines.
-PA-C35966

THE BOSTON MILLS PRESS

This book is dedicated to all those who rush into buildings when everyone else is rushing out – the firefighters.

Canadian Cataloguing in Publication Data

Baird, Donal
The story of firefighting in Canada

Bibliography: p.
ISBN 0-919783-31-7

1. Fire-departments – Canada – History. 2. Fire prevention – Canada – History.
3. Fires – Canada – History. I. Title.

TH9506.A1 B33 1986 363.3'78'0971 C86-093444-6

Edited by Noel Hudson
Designed by John Denison
Special drawings by Audrey Matheson
Typeset by Lexigraf, Tottenham

Published by:
THE BOSTON MILLS PRESS
132 Main Street
Erin, Ontario
N0B 1T0
(519) 833-2407

We wish to acknowledge the financial assistance and encouragement of The Canada Council, the Ontario Arts Council and the Office of the Secretary of State.

Author's Preface

Having passed the age of thirty, I became interested in the history as well as the future of fire protection and, on my travels across the country, began to pry into the background of Canadian fire departments. Resolutions of the Canadian Association of Fire Chiefs and the Association of Canadian Fire Marshals and Fire Commissioners a few years ago expressing the need for preparation of the fire service's history were the stimulus to get busy and put this story together.

The absence of any assembled historical account of the Canadian fire service to date was not only very Canadian, but typical of the fire service, which sometimes calls itself a silent service. There are no heroes widely recognized in this service, but it is full of them. Firefighters, when confronted with their more heroic acts, really do look at their toes and mumble, "It's part of the job."

This project has had the challenge and excitement of covering new ground without previous works to follow. The generosity of many fire chiefs, firefighters, fire buffs, librarians and archivists helped immensely.

There are more than 3000 individual fire department histories behind this book. I have attempted to make them into one, while savouring intimate scenes from individual ones. Some inevitably get more attention, because their stories have been better preserved.

The engine or hose house was a popular rendezvous in the early days where the firemen gathered to socialize and polish up their equipment, such as the graceful spider hose carriage pictured here.

The heroic rescue of servant girl Johanna O'Connor by sub-Lieutenant Beckingham at the St. James Hotel fire in Montreal, 1875.

CONTENTS

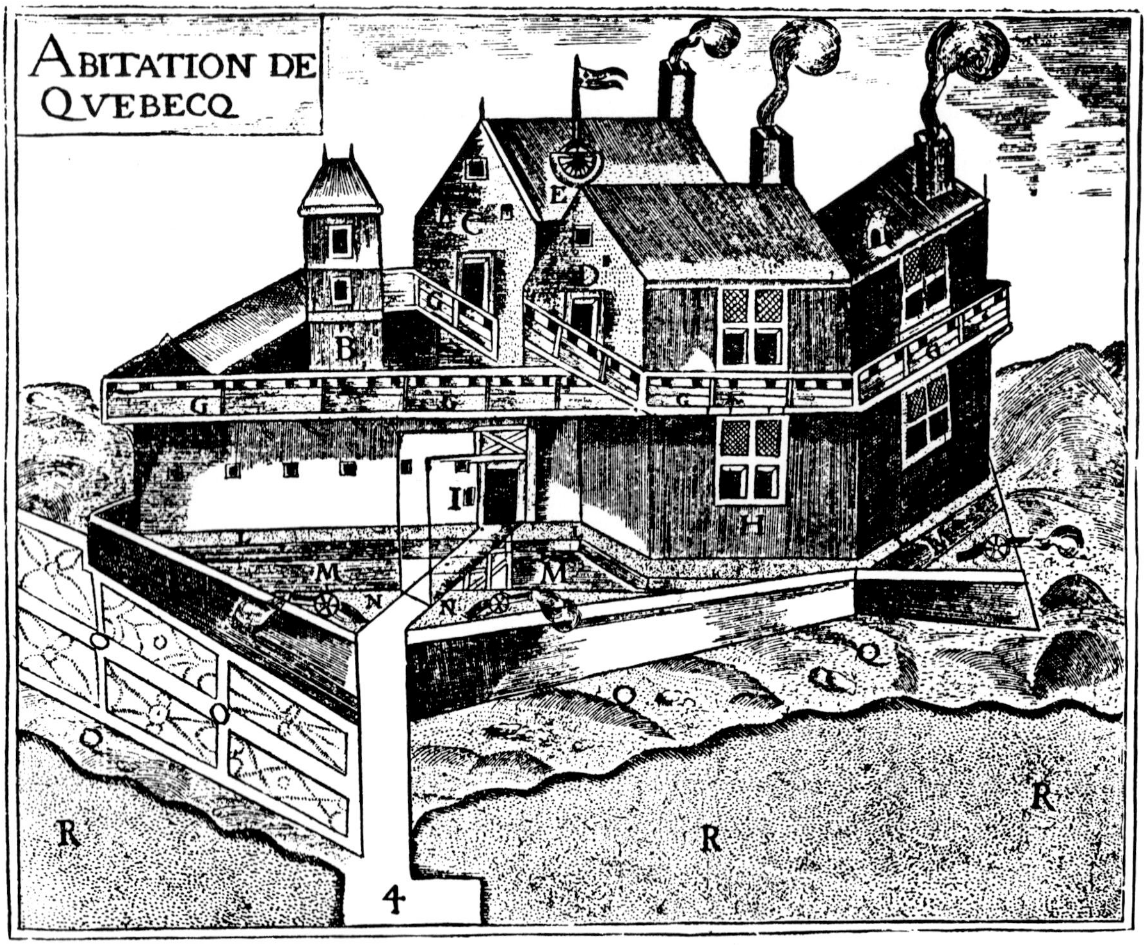

The compact configuration of Champlain's 1609 "Abitation" at the future site of Quebec symbolizes the new country's dependence upon closeknit towns and its vulnerability to fire. - PAC-C9711

1

A NEED FOR PROTECTION (1600-1800)
(I) The Vulnerable Pioneer Community

Samuel de Champlain was doubtless oblivious to what he was starting. When this explorer-settler erected the first few buildings on the site of Quebec, the first permanent town in Canada, he defended them against the Indians with a palisade. But in crowding them snugly together and using the conveniently available wood as his building material, Champlain set a pattern of vulnerability to another enemy. This was fire, whose depredations would always be a salient characteristic in the life of North American cities and towns. The result has been a long history of calamitous fires, and to combat them a strong and colourful firefighting service has been a vital element in the life of every community.

The pioneers who hacked their building sites out of the Canadian wilderness in the seventeenth and early eighteenth centuries, and who hurriedly threw up crude buildings to provide basic shelter, had little time to think about precautions against fires. In fact, not a great deal was being done to this end even back home in Europe.

The firefighting system of the Romans had disappeared in the Dark Ages, and collective protection measures for the community against the scourge of uncontrolled fire were just beginning in some of the old established cities of Europe. In the new cities being started in North America there was a difference. The pioneer towns were, with few exceptions, being built of wood, not stone or brick as in Europe. The colonists were building for disaster, using the material most readily at hand. Expediency overruled any fear of fire and would continue to do so for a long time. The pioneers were to develop a profligate attitude toward the bountiful forests they burned to clear the way for agriculture, and also toward the buildings they built from its wood.

Historical accounts of great fires in Canadian cities and towns give the impression that, after a short period of hardship, the towns were rebuilt and commerce went on as before. Life went on, certainly, and rebuilt towns were often better laid out and better built, but general advancement must have suffered substantially. (This will be seen to be the case in specific instances later..) There is evidence of a tendency to rebuild temporarily in a makeshift way to get basic shelter, so considerable energy was expended before real progress was resumed.

To counter this, there may have been a selective process at work, the crowded, jerry-built buildings being more prone to destruction and then replacement by better structures. This certainly became true after some control over construction was achieved through local building laws and the planning of streets.

Much depended upon the vision, organization and prosperity of the community. Struggling under difficult conditions and with negligible planning or control, St.John's Newfoundland, was burned and rebuilt repeatedly as a crowded tinderbox along its all-important waterfront. As local councils began to take the reins and attempt to develop their communities along controlled lines, generally to bring order out of chaos, a major fire calamity was usually the signal to do something about fire safety.

This action, in the earliest cases, took the form of bylaws prohibiting certain fire ignition hazards, the setting up of night watches, and establishing a measure of control at the scene of fires. The last included prevention of looting, the drafting of bystanders to help, and the provision of buckets and other rudimentary tools for firefighting. Fire engines were available by the mid eighteenth century in a fairly effective form, but they were often only acquired as a second stage of development after the bylaw for fire control. Organized firefighting was, after all, new even in Europe. In some cases it was the British army garrison that showed by example how fires could be extinguished using fire engines, and that merely organizing to salvage furniture and goods from burning buildings was a poor halfway measure.

To accept the inevitability that a building, once it caught fire, would burn down, an absolute loss, and possibly take others with it, was normal through the Dark Ages in Europe and in the new colonial towns in North America until late in the seventeenth century. In 1666 there occurred the landmark Great Fire of London. Out of this were born the fire insurance business and firefighting organizations using fire engines. In Britain the former operated the latter in their own interests.

Before 1700 the Dutch led in development of fire engines, the City of Amsterdam having 60 in use in 1695, an astonishing number by comparison with other countries. The Dutch example gradually spread to England and America. Although there were no engines involved, New York (or as it was then, the Dutch Nieuw Amsterdam) led the New World in establishing a fire prevention organization in 1648. Building inspectors were appointed in that year, with the prevention of fires as their object.

Insurance policies protecting against calamity by fire and the sinking of ships took shape in Britain in the late 1600s, but did not reach Canada until a century later. This meant the homeowner and businessman in pioneer towns took the full shock of fire losses in this era of no fire protection.

Information on fires in the earliest days of Canadian settlement is sketchy, as was the effort to guard against them. An impression may be had, however, of the first blows by fire and the first tentative moves made to reduce such losses in the primitive towns of New France and British North America. Usually learning reluctantly from disaster, citizens and governments slowly implemented protective measures to safeguard their hard-won amenities in the wilds.

The French settled in Canada first, by more than a hundred years. Theirs was a tightly controlled society under a centralized government. It seems to have lacked the expression of a strong local initiative as seen in the British colonies of America, certainly as far as fire protection was concerned.

(II) New France

Champlain's Quebec

To establish a permanent French base, Champlain landed at the site of the future great fortress city of Quebec in 1609 and built three two-storey houses, a protective wall and a watchtower. This was the beginning of the capital of the French regime in Canada. Quebec was the first important town established by Europeans in the Northern half of a largely empty continent. Like many North American cities it was to suffer a long history of great fires. By 1620 this foothold of European civilization, and the first enduring city in Canada, had become a sizable huddle of hastily constructed buildings, houses, storage sheds and wharves, hugging the riverbank among the walnut trees below strategic Cape Diamond. Fishing and farming were beginning to help support the transplanted white men, who otherwise lived precariously at the end of a long transatlantic lifeline.

With the space of a whole continent at their backs, the French settlers built a town in the European tradition, a compact entity designed to secure them from potential enemies and for the social and commercial convenience of their close, interdependent way of living. No nomads like the Indians, they brought along their more complex society, and this needed a town, with its compression of housing, services and goods in a small area. Towns were marked by their accumulation of worldly goods, buildings and their contents. By their nature these were vulnerable to destruction by fire.

Trading posts and settlements were established up and down the St. Lawrence River, the great highway into the interior of the continent. There was nearby Sillery, the more distant Trois Rivieres and Tadoussac, remote links far in the interior, and the one that would surpass Quebec to become the commercial metropolis, Montreal. Sometimes these were lawless freetraders' bases like Tadoussac, where fur traders and fishermen established temporary shack communities as shore bases for a season, then sailed home. The permanent settlements with their supporting farms outlying them, with ingenuity and labour, performed surprisingly well in providing at least the bare necessities of civilized life for the transplanted european in a somewhat hostile environment.

But the greater the luxury, the greater the loss if fire struck. The Indians, if they ever set fire to a tepee, recovered much more quickly than a French official who had to wait for a Spring supply ship to replace his losses. The New World differed from France in its severe climate, the need for heating in winter and a great dependence on wood for construction. When a fine home or an important facility was destroyed in early Quebec, there were not plenty more to draw upon, nor were there supplies of furnishings and tradesmen available to quickly replace that lost. Destruction of a building and its contents would have been a hardship, and when fire spread through a substantial part of a community, the setback must have been severe.

A painting represents one of the earliest fires on record in Canada, the burning of the first monastery in Quebec in 1650. Indians watch as the white man's shelter from the wintry wilderness is consumed. - PAC-C117818

The first recorded fire of note came during Champlain's tenure as governor when, in 1629, an incendiary fire is said to have been set by the brothers Kirk, agents of the British, who occupied Quebec briefly while under orders to capture the colony. Champlain was not oblivious to the danger of fire within his palisade and took steps to fight it. It is reported that he ordered the tolling of the bell in the tower of his "Abitation de Quebec" as a fire alarm, the soldiers and citizens to turn out with axes, bars and buckets. Aside from what could be done with bucketsful of water, the main effort was through the demolition of buildings in the path of the flames. This was a forward step, organized firefighting being a rare thing anywhere at that time.

Under Governor Charles de Montmagny, in 1640, some of Quebec's flimsy structures were replaced by stone buildings. These were most often schools, churches, hospitals and commercial buildings, buildings were climbing the hill from the Lower Town shoreline to the summit of Mountain Hill. The population was all of 200. The well-to-do lived much in the elegance of France, provided with the luxuries of home by the occasional supply vessel.

Montmagny set about improving the fort and habitation of Quebec with vigour, but his proud new Champlain Chapel and the adjoining Jesuit House succumbed to flames after only a short existence. It had been a major achievement of construction in a rough settlement, built proudly as a mark of Christian civilization among the savages.

By 1682 the main part of town, below the Cape, consisted of about 80 buildings. At this point luck took a turn for the worse. The first of a number of conflagrations struck the growing capital of New France. On the sixth of August fire broke out in the Lower Town and the price of hasty, crowded building in flimsy wood was paid, as it would be again and again.

Attempts at arresting the spread of the fire consisted of tearing down the lightly-built structures in the path of the flames, using axes, saws and ropes, as there were no fire engines. In spite of desperate efforts the conflagration swept through two-thirds of the crowded business area while the merchants and the governor were away in Montreal trading. This was the first conflagration on record in Canada.

Under Siege

Destruction of cities by enemy action is rare in Canadian experience, St. John's by the French, Toronto and Niagara by the Americans being worthy of note. But the most destructive was the bombardment and firing of Quebec by the British under General Wolfe. When the British launched the campaign that finally ended their fight with the French for the control of British North America in 1759, they lay siege to both Louisbourg and Quebec, in each case bombarding them with naval guns.

At Louisburg the wooden homes of the poor were burned in the attack, but the demoralized fortress was doomed regardless of any effect of this. At Quebec the bombardment was intended to cause the inhabitants to abandon the fortified city, and it went on intermittently in July and August. Six 32-pounders and five 13-inch naval mortars were utilized by Wolfe's forces, shooting solid shot, shells and incendiary carcasses.

The church of Notre Dame des Victoires in Lower Town and the cathedral in Upper Town were burned, along with 180 houses, warehouses and commercial establishments, about half the town. First to go was the seminary, exposed as it was on the heights of Upper Town. It had already burned down and been rebuilt due to the carelessness of its occupants in 1701 and 1705. This time, on July 16, five English mortar bombs landed on it and more bombs were fired to discourage firefighting.

On the ninth of August, a great quantity of firepots and carcasses was dropped on Lower Town by mortar from across the river. With winds blowing from the southwest, shifting to the northwest, the fires the bombs started were impossible to hold and a general conflagration ensued, destroying 150 buildings. The strategic value of this destruction has been questioned. It was hard on civilian and military morale, but did not affect the fortifications much.

The effective combination of explosive and incendiary shells was well known to the British and used against the French strongholds of Louisbourg and Quebec. Much of Quebec was destroyed in August 1759 before its capture. Pocked and roofless walls of the Cathedral and Recollect Church remain. - PAC-C361

There are historical references to firemen among the defenders of Quebec, but these were in all probability military fire *picquets* or other personnel assigned on an emergency basis. It is conceivable that the French garrison troops had fire engines at this time, though there is nothing to confirm this and it must be concluded that there were none.

Outpost Montreal

Montreal was in these times one of the young satellite settlements of the capital. It was founded by Sieur de Maisonneuve as a commercial trading post in 1642. Named Ville Marie de Montreal, it was located on Montreal Island, at the junction of the St. Lawrence and Ottawa Rivers, a marvelously strategic location for the fur trade. The Indians also valued it as a transportation crossroads. Their animosity over the intrusion of the French required that Maisonneuve's habitation be fortified with a palisade, and when houses were built outside the fortified area, courage and watchfulness were essential.

Starting with 60 people, birchbark tents and a log palisade, the new hub for the fur trade grew rapidly because of its undoubted commercial appeal and the resources pumped into it by its sponsors back in France. Along with the business enterprise was the mission of Jeanne Mance to establish a convent hospital on this savage frontier. The first fire recorded in Montreal occurred in 1651, when two houses were burned by a party of forty Iroquois. The second was the loss of the convent in a fire that took the lives of two nuns in 1683.

Although Montreal was thriving and growing apace, reaching 766 inhabitants in 1667, it was oppressed by the constant menace of the Iroquois. The Indian hit-and-run attacks hit a peak in a most barbaric case of arson and murder two years after this. On August fifth, there occurred an example of the use of fire in war at its most hideous. At dawn on that day a war party of 1,400 Iroquois descended upon the little settlement of Lachine, at the rapids west of Montreal. They surprised the inhabitants in their beds and proceeded to sack and burn the homes, burning 200 of the helpless colonists alive and dragging off many others to be tortured, often by fire. The homes, crops and plantations were all reduced to ashes.

The Montreal of the French colonial era suffered two large-scale fires before taking any measures to fight them on an organized basis. The first, on June 19, 1721, started during the religious procession of *Fete Dieu.* During the firing of a salute, it is recorded, a soldier "imprudently fired his salute towards the cornice of the Hotel Dieu, setting it on fire." The flames spread from this nuns' hospital to raze some 138 buildings within the crowded walled part of the town. It was the worst of the early fires in Montreal.

The second conflagration was a case of arson on April tenth, 1734, being set by a disgruntled slave as an act of revenge involving an affair of the heart.This time 46 homes were consumed, along with the rebuilt Hotel Dieu. The fire was significant because it stirred the authorities to set up a firefighting organization, the first in what is now Canada, as far as we can tell.

Yet another conflagration occurred early in the British era. On May 18, 1765, a fire destroyed 108 houses in a few hours, displacing 215 families. It burned the market area and along St. Paul, St. Francois, St. Louis, and St. Sacrement streets, crossing the city wall and St. Peter's River to spread into the westerly St. Anne's Suburbs. It was stopped by pulling down part of the Grey Nuns' Hospital and some homes near it. Relief funds were raised in England, King George III giving £500 toward them. It was reported that the French-speaking citizens were surprised at the British rush to provide relief to such a far-off colonial outpost.

This fire started in the home of a man named Livingstone when hot wood ashes were carried into the attic for soapmaking. The extent of the fire was attributed to a prevalence of high winds and there being no fire engines. The lesson was heeded, the authorities taking steps to acquire engines for both Montreal and Quebec.

Fire Regulations In Quebec

As may be expected, the first indication of community fire protection measures being initiated in Canada occurred in the first town of any consequence to exist. This was Quebec, capital and first city of New France. References in accounts of the seventeenth century show that there was a 1692 regulation in Quebec requiring the householder to clean out his chimney at prescribed intervals. Shortly after this, another ordinance required every ablebodied person to respond to fires with an axe and a saw.

At that time the only firefighting was with buckets of water on a very much ad hoc basis by citizens at large and with tradesmen tearing buildings apart using saws, axes and poles with hooks. At the major fire in August 1682, in Lower Town, all serious efforts to arrest its progress appear to have taken this form.

Such water as was thrown on fires was carried from the St. Lawrence and St. Charles rivers or wells in Lower Town. A well was dug in Upper Town expressly for fire purposes, because the rivers were too far away to be of use. The traditional bucket brigade — lines of people passing full and empty buckets back and forth — was used, but as a rule, once a building was on fire complete destruction followed.

Better technology was not entirely unknown in New France, however, as the Sovereign Council in 1691 granted permission to inhabitants of Lower Town "to import from France a fire engine of the Dutch style to throw water on houses in case of fire." As noted, the Dutch did have superior fire engines at the time, being the leaders in the field. Nevertheless, the cryptic comment is also recorded that the engine never arrived in Quebec. Why, is not explained, but it is worth noting that it seems to have been the citizens, likely the major property owners in the business quarter, who took the first initiative to buy a fire engine.

The Quebec Seminary burned down in 1701 while the clerics and students were away on an outing in the country, consuming the work of 30 years in a mere hour and a quarter. Following this disaster in the Upper Town, the authorities were said to have "wakened up a little." An engineer proposed and dug a new, deeper Royal Well to replace the old one for the protection of the buildings in the central part of Upper Town. There were 50 leather buckets distributed about the whole town when, in 1706, the Council Royal had another 100 made for distribution.

More attempts to control hazards were made in 1726, when cleaning of chimneys every two months was called for. The householder was required to have this witnessed on paper by two neighbours. Wood fires, creosote-laden chimneys and wood shingle-covered roofs were always recognized fire risks in the early days, as will be observed in the early fire precautions in most towns.

The year after this, the Intendant Dupuy prescribed a construction ordinance forbidding the erection of buildings in the towns of the colony that were more than two storeys, were not built of stone or were roofed with wood shingles. A commentary states that Dupuy left New France two years later and his ordinance "became a dead letter, or nearly so. His law would have contributed to better protection of Quebec towns but failed as being too soon in a world too young."

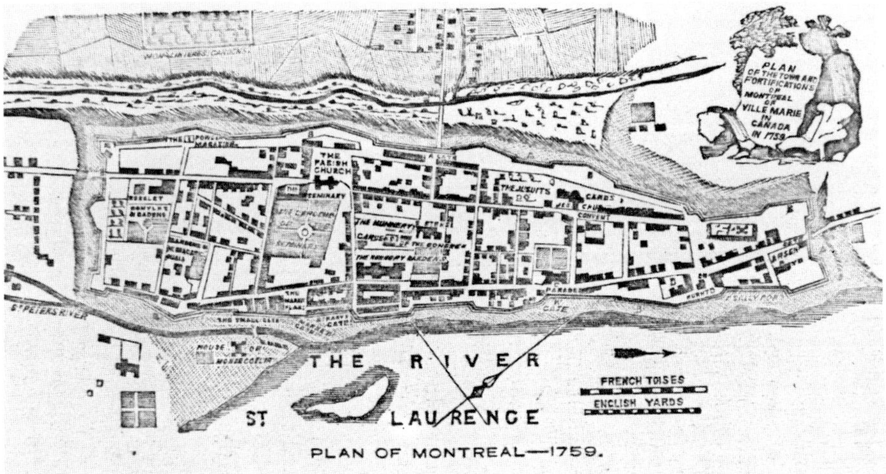

Montreal was built as a walled city, predominantly of stone. This compact fortress remained as the downtown commercial centre after the city spread rapidly beyond the walls in fire prone wooden suburbs.

A Firefighting Squad In Montreal

In view of the destructive conflagrations of 1721 and 1734 in Montreal, the Intendant of New France, Gilles Hocquart, conferred with the civil governor of Montreal, the Marquis de Beauharnois, his military counterpart, Lieutenant General Beaucourt and the local merchants. The result was an ordinance dated July 12, 1734, calling for the distribution of tools and utensils for firefighting in the town. These were to be marked with the state fleur de lis symbol and to include 80 leather and 200 wooden buckets, 100 axes, 24 hooks, 12 ladders, 100 shovels and 12 demolition tools. This equipment was to be located at the Jesuits' Church, the Guard House, the market, the Recollets' Church and the seminary.

The ordinance also provided for the enforcement of chimney cleaning rules. This was an important matter since the slow-burning wood fires deposited creosote in the chimneys. The creosote sometimes caught fire, spreading sparks about the neighbourhood, with its wood-shingle roofs, as well as passing through defects in the chimneys to the wood of the house. Some of the early chimneys were of clay daubed on wood, subject to rapid deterioration.

The procuring and distribution of such firefighting tools was to be a common first step in municipal fire protection measures in the early days. It was usually intended that the whole populace turn out to use them. In the case of Montreal, however, Hocquart ordered that "the carpenters, masons and roofers and other workers most necessary at fires be set up in two squads, each to be under a commander who shall issue the orders." A fine of six livres was provided for failure of these tradesmen to comply. This appears to be the first recorded organization of a designated firefighting crew in Canada.

This first firefighting service must have been effective, at least in the eyes of the authorities, as four years later Louis Truteau, a 47-year-old carpenter, one of the commanders, was recognized "for his zeal and attention that he particularly showed in the different fires that occurred in Montreal." Truteau was awarded a gratuity of 30 livres a year. In the Intendant's ordinance announcing this, Truteau was also charged with the duty of inspecting the buckets and tools every 15 days if needed, but not less than once a month.

The French regime in Canada was not such as to inspire local initiatives like community fire protection. In any case, it was only in the latter part of the eighteenth century that the example of the value in self-starting fire protection by clubs, associations or local governments was demonstrated. It was to come largely from the self-reliant New England colonies, often brought in person by the loyalists.

(III) The British Regime

The First Fire Engines Appear

A *Club de Feu* or fire club was formed in Montreal in 1763 under the new British regime. It was typical of the early organization also known as a protection club and was an association of propertied merchants and prominent citizens. Having the most at stake in the event of fire, they banded together to assist one another and the community. Their main effort was toward the rescue or salvage of goods from endangered buildings, with some passive efforts at preventing fire spread. At this time in Montreal the garrison soldiers were more prominent in any actual attempts to extinguish fires.

A careful distinction needs to be made between the protection club or association, also called a fire club or fire company at times, and the fire engine company. The former might use buckets and hooks to fight fires, along with its salvage mission, but the latter performed progressively more effective and aggressive battles to extinguish fires. Little more than a large wooden box on wheels with a pump set inside it, the fire engine could direct a continuous stream of water as high as the roof of most buildings or well into the interior through a window. Small as it was, the engine's nozzle stream utilized limited water much more effectively than dashes with buckets.

Descriptions of the conflagration of 1765 in Montreal say that there was "a want of fire engines." The muscle-powered pumping engines, whose box or "tub" was filled by buckets, were beginning to show their great value in extinguishing fires elsewhere at this point, notably in the New England and Southern colonies. Following the big fire, the authorities ordered two English pumps, and practices were held every Monday. Soon, two more pumping engines allowed for one to be provided in each of the four quarters of town.

At that time water was obtained mainly from the public well in the central square of Place d'Armes. A Sergeant Taylor was appointed the first fire chief.

Early firefighting companies, like many volunteer organizations, often must have lost their enthusiasm, for they were frequently being reshuffled and reconstituted. Some early accounts represent the first fire department in Montreal as being that set up in 1786. In that year a meeting was held in the Convent of the Recollet Church by order of the magistrates. A committee of 13 English and 13 French citizens was appointed to draft regulations for a fire department. A member of the committee was Alexander Auldjo, who was to become the first fire insurance agent in pre-confederation Canada.

There could only have been some dissatisfaction with the existing firefighting arrangements. The committee organized things on a ward basis, with an engine in each. This seems to have produced the desired results, as there was little news of change on the Montreal fire protection scene until the period of rapid growth after the turn of the century.

Downriver at Quebec, two fire engines were also obtained in 1765. One must suspect that this was not coincidence, but evidence of the government having a hand in deciding both major cities should have a couple of these machines. The Quebec *Gazette* of June 20 stated, "two pumps and forty feet of hose arrived" and "The subscribers to the fire engine are desired to meet this evening at five o'clock at the house of John King in the Lower Town, to fix a proper place to lodge it in, appoint persons to keep it in order, and other matters relating thereto."

A volunteer fire company was formed under government charter in the following year, known as the Amicable Society for the Prevention of Fires. It is not clear where it stood in the fire protection scheme. It was possibly concerned with the areas of property salvage or mutual insurance and could have been the predecessor to the pioneering Quebec Fire Insurance Company.

This insurance company operated its own fire engine company, the only one to do this in Canada, apparently. There have been suggestions that other companies operated firefighting units in Quebec, but it is generally the case that supposed insurance fire brigades in Canada were a variety of financial and other encouragements to existing companies.

On April 11, 1768, now under British rule, Quebec suffered another in its succession of major fires. An outbreak in a stable in Upper Town started at five p.m., spreading through 90 houses, two churches and a large charity school, lasting all night. Although they could not do much with a severe fire once it got away, there were now two hand-powered fire engines available, the beginnings of firefighting in the city.

Honours for first firefighting company equipped with a fire engine in Canada would have to go either Montreal or Quebec, probably the former, the date being 1765. However, there is a lack of specific detail that would confirm this with any precision. The Montreal Fire Service dates its origins from 1842, much later, with the uniting of the various fire engine companies into the city fire department that has continued down to today.

A Garrison Town

A third city of strategic importance in this era was the one the British built in a hurry, the fortress of Halifax. Founded in 1749 by an army-navy expedition, it grew rapidly under the impetus of the campaign to eliminate France as a power in Canada. Representative government for the Province of Nova Scotia was granted as early as 1758. In its first year Halifax had 300 houses erected where previously there had been wilderness.

Halifax was founded by Hon. Edward Cornwallis, who came from England in 1749 with a fleet of warships and transports carrying 2,400 people to build a fortress countering the French Louisbourg up the coast. Like the French cities Halifax was a garrison town with the army and navy prominent in its affairs. Eight years later Admiral Holborne arrived with 6,000 troops, and Amherst and Wolfe came in 1759 with 159 ships and 12,000 men, preparatory to the siege of Louisbourg. The town swelled rapidly, with tent cities handling the overflow when large contingents arrived.

The civilian population expanded and the merchants prospered, but not before the demon fire got in a good blow. While the town was still mud, stumps and buildings under construction, on July 11, 1750, it was severely set back by a conflagration. The Halifax authorities were thus made conscious of the threat of fire and took some preventive measures within three years of the first settlement, even though this city was more fortunate than most in being free from major conflagrations, after its initial experience. The British

Halifax was hacked out of the woods and rocky hillside as a planned town and military bastion. Fire visited it, but good planning reduced its effects. Wooden St. Paul's Church and the Parade in 1764, both still in use today. - PAC-C4293

military presence may be the explanation for this in part. The first law for the control of fires in Halifax was enacted in 1852, although the first fire engine company would not appear for another 34 years, just shortly after the Loyalists landed.

Ten years later, the Nova Scotia House of Assembly enacted new legislation appointing ten fire wards and providing that persons convicted of stealing at the time of fires were "to be judged felons and shall suffer death, as in cases of felony, without benefit of clergy." An act was also passed in 1762 to prevent the firing of squibs, rockets, serpents or other fireworks, but it is to be hoped without such severe penalties.

Not waiting for any further great fire to demonstrate the need, the Governor in Council in 1752 enacted a law "providing in case of fire for the more speedy extinguishing thereof and for the preserving of goods endangered thereby." It provided for the appointment of fire wards in Halifax having, in conjunction with the magistrates, powers to conscript citizens and to pull down, blow up and extinguish buildings having fires. Buildings not on fire could be pulled down to stop the spread of flames and a tax was provided to pay compensation for the owners. According to Halifax council minutes of September 29, 1757, wardens were to carry six-foot red staffs with brass spears as their identification and, possibly, to intimidate those they were ordering about.

The first firefighting company of record in Halifax was the Union Engine Company, founded August 8, 1786. An article in the Halifax *Herald* of December 21, 1897, refers to the Union Engine Company "holding a rediscovered record of a fire department which existed from 1754 to 1759." No more is recorded on this possibility.

A "Fire Company" was established in Halifax on January 28, 1789, carrying the name previously used in Philadelphia by Benjamin Franklin's company, Hand in Hand, raising the

The first attempts to establish a measure of fire control frequently saw the appointment of fire wards or wardens to keep order, prevent looting and organize bystanders to help.

speculation that it, too, was conceived as a mutual insurance and protection organization. The Nova Scotia *Gazette* carried an announcement in which the governor requested that the gentlemen he had called upon to form a fire company meet to agree to rules and regulations.

The hand in hand men were equipped with salvage bags as their chief tool. A second such company came into being in 1801, carrying the name Sun. It was to become a prestige club, with membership from among the most prominent merchant families. These two companies were later joined by two more, the Phoenix and the Heart in Hand, making for an unusual number of protection or salvage companies in one city.

The protection companies supplied their own bags, baskets, and other tools devoted to the rescue of goods. The city provided buckets, ladders and, presumably, fire engines. As early as 1783 the city was threatening prosecution if town buckets missing after a fire were not located. In addition to buckets for filling the earlier fire engines, they maintained portable well pumps for filling the buckets and built tanks or cisterns at strategic locations. The harbour wharves were also a handy source of supply for the bucket brigade lines.

By 1810 there were at least two fire engines and an engine house in Halifax. Unlike other places, where there were a number of individual companies each independently manning an engine, each Halifax fire engine was in the care of a "section" of the one Union Engine Company. The sections were added as the fire stations grew in number with the spreading city. Ladder and access services were provided for them by the Axe Company, commencing in 1813.

Instructing his company's agents in Halifax to reduce fire insurance premiums following an inspection tour, the actuary and inspector of the Phoenix Assurance Company, Jenkin

Jones, on July seventh 1808,referred to "an abundance of engines and the promptness of the fire companies." Part of the strength of response he was referring to was certainly provided by the army garrison and naval dockyard establishment. Jones reported to his directors on the Halifax of 1808 as a town where brick buildings with slated roofs were gaining ground. There were occasional breaks in the continuity of the buildings along the waterfront. It had a regular plan, he said, with streets of good width and the buildings in the "suburbs" or extremities well dispersed. There was the added feature in its fire defenses of the many military sentries stationed about the town. These transmitted alarms rapidly by bugle.

Dartmouth, Nova Scotia, started a fire department in 1822 with the organization of its own Union Engine Company. Before that, sections of the Halifax U.E.C. had crossed by the harbour ferry on occasion to fight Dartmouth fires, but the delay in transit was not conducive to fire control. A good example of early mutual aid cooperation is the expedition from Halifax when Stanford's Tannery burned in July 1844. A ferry was requisitioned for an engine from the Halifax company and "one ably conducted by a party from the Rifle Brigade" to render assistance to the Dartmouth company.

The Dartmouth company was organized September 21,1822, under Captain William Allen with a mere nine men at the beginning. There was a board of fire wards, an engine and a new engine house on Queen Street at Wentworth. The company met once a month at sundown. Chores, like cleaning snow from in front of the doors, were assigned by drawing lots.

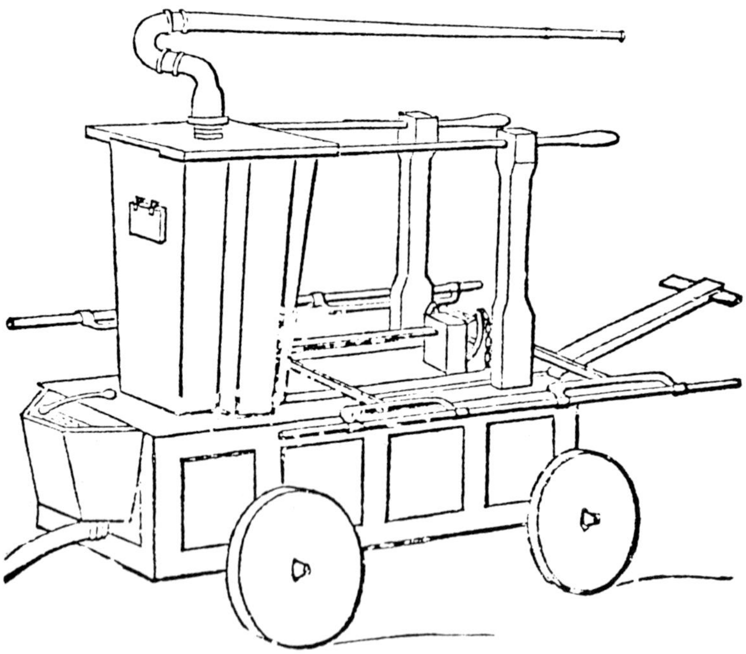

A landmark in the development of the hand fire engine was that of Richard Newsham of London. Many of the early Canadian and American engines in the late 18th Century were of this pattern.

The Loyalist Towns

After the middle of the eighteenth century the British began to build numerous towns in Canada. After the American War of Independence, thousands of displaced Loyalists came north to swell the small settlements and found new ones overnight. Some, like Saint John, became important cities, but others, like Shelburne, did not sustain their initially promising impetus.

Although the Loyalists had to struggle to create even the rudiments of civilized amenities in their raw communities, they brought with them a significant and useful idea, the independent volunteer fire company. Coming from the well-established urban areas around Boston, New York and Philadelphia, many of them were accustomed to having fire protection. Firefighting with fire engines dated back as far as 1697 in Boston. In fire protection, as in other things, the way of the colonial was to get together with his neighbours and do it himself. This would be the great distinctive feature of the fire service in North America.

While some cities have had very little of their early fire protection history preserved to the present, others fortunately can give us a fairly full picture of the struggle to achieve a measure of protection against the depredations of fire. Saint John New Brunswick, is one of the latter, and probably not just coincidentally, is a city with a history of numerous great fires.

Saint John was no more fortunate with fire in its first year of existence than Halifax. Twenty transports carrying 3,000 Loyalists descended upon the tiny settlement of Parrtown, at the mouth of the Saint John River, in the Spring of 1783. They were met by Major Studholme of Fort Howe, but there had been virtually no preparation for them. Many of these Loyalists were upper-class and professional people from New England. However, all set to with determination. Streets were laid out and huts thrown up on the hillsides to provide shelter. Later in the year another 7,000 arrived, about half of these moving on to settle elsewhere.

Musclepower of the citizens supplied the water at fires until well on in the 19th Century. Many of the earlier engines had no hose so the engine had to be placed right in front of the fire.

Fire struck the spartan habitations of Saint John in the following year, burning many buildings in what could only be described as a major reverse in the building of a town. Certainly the buildings lost would have been small and rough, but construction of replacements would have been fraught with shortages of materials and labour. As well, the burned homes would have held irreplaceable family effects brought in the exodus from New England; a discouraging loss.

Nevertheless, Saint John became, by Royal Charter in 1785, the first city to be incorporated in Canada. The Common Council first met in May, by October had passed a fire prevention bylaw, and on December 30 ordered an enquiry into where and at what price fire engines could be obtained. It was reported back by January second that two were available for £136 Sterling. They were recorded as purchased within the week from one Peter Fleming, and it was decided to advertise for "Proposals for erecting two companies of firemen." One was to be for the Upper Cove area and one for the Lower Cove. This was fast progress, but then came the setbacks.

It is easy to conclude that conflagrations that so soon swept the hurriedly built towns of Halifax and Saint John were simply the direct result of flimsy, inferior construction. However, when forests are cleared to build houses there is an accumulation of brush (called slash) that becomes highly combustible when dry, and this is likely to have been an important element, at least in the case of Saint John.

The primitive engines bought by Saint John proved to be completely unsatisfactory. Perhaps they were secondhand or homemade imitations. In any case the city had made a bad deal and had no money to replace them. In addition, the invitation for proposals to provide fire companies developed into protracted negotiations over the special privileges granted to firemen. There was an immediate proposal by fourteen men to form a fire company for Upper Cove (the present downtown core area), but in another year, in which council arranged for the erection of an engine house in each of the Upper and Lower Cove marketplaces, there were recruiting problems reported by the two embryonic fire companies.

The two appointed foremen for the companies, Abel Hardenbrook and Adam Henneger, said they "find it impractical to procure proper persons to engage as firemen to work the said engines unless they are exempted from serving on juries, besides the other privileges granted them by the Act of Assembly." They asked the mayor to use his influence to gain this privilege, these being within the powers of the Provincial Assembly. Dickering over the exemptions — which became a widespread privilege for volunteer firemen in Canada and the United States — continued in Saint John until 1792, when it was settled that Saint John firemen should be exempt from all jury duty except the Supreme Court, from serving as constables, militia duty and obligatory street work except before their own doors.

In the meantime, a chimney cleaning bylaw, appointment of a public chimney sweep, provision of ladders, pulldown hooks, leather buckets, and the selection of locations for eight wells had been seen to. The purchase of four dozen leather buckets for £18 was arranged through a barter deal involving sale of a city lot to a third party.

Major General Benedict Arnold contributed toward the purchase of this first fire engine in Saint John, shown to be a primitive gooseneck type in this sketch made when long abandoned.

To solve the problem of replacing the lemon fire engines, in March 1787 the two fire companies proposed to pay for two new engines and to dig wells for them if the city would repay them within two years. The council replied that they "are fully sensible of the public spirit discovered in the proposal, but in the present embarrassed state of the city funds, cannot undertake to repay the sum within the time proposed." They offered to pay the interest only until the coffers were in better shape. Leading citizens came to the rescue, pledging donations to pay for the engines. Prominent among the pledges was £10 from Major-General Benedict Arnold (retired).

A fire club was formed in Saint John, apparently in 1795, called No.1 and later Wellington Fire Club. But there was a night fire watch set up in 1790 by the city said to be in response to a request from the fire club. Perhaps it was an earlier effort that did not survive. The Wellington Club became Wellington No. 1 Engine Company, occupying the same Market House location on Brittain Street as the original Lower Cove engine company. Possibly all three entities were one and the same unit in different incarnations.

The night fire watch consisted of four men contracted to patrol the streets from eight p.m. until sunrise, sounding fire alarms, picking up disorderly persons and similar duties.

Seemingly preoccupied with chimney fire problems, the Saint John Council in the same year again revised its bylaw for chimney cleaning and decided that the aldermen could license the inhabitants of their wards to burn out their own chimneys instead of having them swept — provided it was done between sunrise and noon, and in rain or snowy weather because of the spark danger.

Two more engine companies and a hose-and-ladder company were added up to 1820, while the first overall chief, David Waterbury a Loyalist from Connecticut, was appointed in 1816. This formalized a municipally controlled and coordinated firefighting force.

David Waterbury was first over all fire chief in Saint John. A gentleman Loyalist from Connecticut, he would have been well acquainted with the work of established fire companies in New England.

The new towns courageously carved from the forests had their ups and downs, but succeeded in the long term in spite of being built to burn. One that did not was the first town of Guysborough, founded at Gambier Bay, N.S., by one of the disbanding British regiments. The settling soldiers arrived in 1783 to build a town of some 300 buildings and a population of 800. Soon a furious conflagration ripped through the settlement, driving the inhabitants into the water, destroying it almost entirely. Considering the apparent ferocity of the fire, it is doubtful fire engines could have stopped it had there been any.

Property and livestock were gone and the people left without shelter, bread or tools. Famine would have ensued but for a King's relief ship dispatched from Halifax. Instead of picking themselves up from the ruins and immediately rebuilding as most communities would, the Guysborough people reconsidered. Their site had proved to have indifferent harbour shelter and the soil was poor for farming. Most of them made the decision to move to Chedabucto Bay, where a new Guysborough was born.

With Guysborough the exception that emphasizes the rule, the populace of a community largely wiped out by a conflagration usually rebuilds on the same spot. This seems to confirm that there was usually adequate reason for the town to have been located were it was. Other reasons are the existence of streets and facilities such as water pipes that are still there, to say nothing of the lots of land that are still owned by those who must rebuild somewhere.

Shelburne, N.S., was settled by a well-organized expedition of Loyalists, 471 families arriving in 18 ships in 1783. An orderly townsite was laid out, but a horde of additional and not particularly welcome refugees flooded the town in the next few years, boosting the population to 12,000 in 1784. It became a town of 3,000 houses, half again as many as in Saint John, a similar Loyalist town.

Shelburne had been long credited as receiving the first fire engine in Canada, in 1786, and this as a gift to the settlers for their loyalty, many of them demobilized soldiers, from King George III. It certainly was not the first fire engine and there is considerable doubt that it was paid for by other than the citizens themselves. Possibly its arrival aboard a British naval vessel confused the record.

Shelburne was spared any great fires in those early days, but being off the beaten track and in spite of its magnificent harbour, declined and lost out to other towns in importance.

Just after the turn of the century, Fredericton, another Loyalist town in New Brunswick, had the benefit of an army garrison engine running to town fires. Made a city when the size of a small village because it was the site of the Anglican cathedral, fire wards were appointed in 1817 by the governor. They were to be "eight prudent and decent persons from the gentry and commonality of the town." Each was to be equipped with a seven-foot red staff and a speaking trumpet. In addition to keeping order at fires, they were to inspect fireplaces. The act included the usual fire prevention cautions like forbidding of persons to enter barns with an open candle. A volunteer fire company was set up to take over from the army's courtesy service in 1820.

This drawing from the London Illustrated News suggests that by the 1880s the Fredericton Fire Department was able to make an organized attack when the Provincial legislative building caught fire.

- PAC-C72867

The Miramichi Fire Tongue

Calamity befell Fredericton in 1825 in a fire quite beyond any fire department's powers of resistance. In a summer marked by heat, two months of drought,and forest fires all across North America, a great fast-moving forest conflagration swept a swath across north central New Brunswick. An extension or offshoot of this furious "Miramichi Fire Tongue" — which had started in the lumbering country along the river of that name, about 100 miles to the east, a couple of days earlier — it cut through the centre of Fredericton on October seventh. It destroyed much of the town, including Government House.

One of the great forest holocausts, in a class with the much later American Peshtigo and Canadian Haileybury fires, the Miramichi fire travelled from the mouth of the river on the Gulf of St. Lawrence, across New Brunswick to the Bay of Chaleur, leaving over 5,000 square miles of desolation. Like the other holocausts, it was characterized by great winds and incredible speed. For many farmers who had cleared farms in the depth of the great central New Brunswick forest, there was no escape.

The effect of this fire upon the thriving lumbering towns on the lower reaches of the Miramichi was much more devastating than on Fredericton. The navigable river and extensive stands of prime timber in the interior made them very important commercially in the early part of the century. The villages and commerce on the river grew rapidly.

On the fifth of October it was hot in the area of Newcastle, the largest town at 1,000 souls and 260 buildings. It was hazy, with thunder and lightning in the distance. In the late afternoon the heat grew more oppressive and a dark pillar of smoke was seen approaching. The wind came up, increasing in strength.

The people in Newcastle and nearby Douglastown gradually became more and more alarmed. By nine o'clock, the wind was blowing a gale, churning the river surface. Then a storm of fire came hissing through the forest, sweeping down upon the settlements, rapidly reducing houses, mills, stores and barns to ashes. It rolled through the towns with little or no warning and the same rushing fury as the later fires like Peshtigo and Haileybury.

Most of the inhabitants hurried to the nearest refuge, the river, plunging in up to their necks. Some got into boats or on rafts of logs, but they were driven by the wind into the storming river to drown. Some fled to the marsh southwest of Newcastle and were saved by the lack of fuel between them and the fire. Sailors fought to save their vessels, fighting the fire catching in the rigging, but three ships were destroyed.

The conflagration extended right across the north central part of the province, reducing Newcastle from 260 to 12 buildings and destroying Douglastown, as well as smaller villages. Some 160 people were burned to death or drowned. There had been four thriving towns along the river in a well-settled hundred-mile stretch. Warehouses and barns were full for the coming winter. The area was left destitute, with resupply almost impossible so late in the season.

(IV) First Towns In Canada West

The Americans Burn Niagara

Through the first half of the nineteenth century, while the older centres of the Maritimes, Quebec and Newfoundland were reorganizing and expanding previously established fire protection measures, the emergence of towns and fire protection for them was just beginning in Ontario.

Small towns like York, Hamilton, Kingston and London began to grow significantly as the tide of immigrants from the British Isles and Europe pushed inland during this period.

The towns grew in response to commercial and social demands and, as the vulnerability to fire also grew, the citizens looked for fire protection along with other civilized amenities and services.

Public fire protection was no longer something new in the world, as it had been a hundred years before. It was becoming a demonstrated system, with previous examples set in the Canadian and American east.

Long before Simcoe founded Toronto (or York, as it was then) as the new capital for Upper Canada, the old capital was the established town of Newark, or Niagara, across the end of Lake Ontario, where the military base of Fort William stood at the entrance to the Niagara River. In the early years of the nineteenth century it was developing into the commercial distribution centre for what was then Western Canada.

On the opening of the War of 1812, the invading Americans shelled the town. The local militiamen collected the citizens' buckets and fought the fires that ensued. The jail, courthouse, chief engineer's quarters and some houses were soon wrapped in flames from the hot shells in this October attack. The main barracks and stores were saved, while the roof of the magazine caught fire and was daringly extinguished. There is no mention of a fire engine.

The occupation of Newark by the Americans drew to a close on December 13. Their commander, General McClure, had been repulsed by the British and retired upon Fort George for the winter, but was ordered by the U.S. Secretary of State to burn the town and retire across the Niagara River. As writer Pierre Burton has said, it was a war noted for its acts of arson.

American soldiers burn the town of Niagara during the war of 1812 while the men were away with the Canadian Militia.

Newark was a well-built town, with many large residences of brick or stone. There were about 300 houses and commercial buildings, a centre of trade, military base and seat of government. There were approximately 1,500 inhabitants before the war, with 600 to 1,000 troops in garrison. Buildings included a library, two churches, schools, the government house and a number of hotels.

After allowing the householders an hour's warning to remove their belongings, the American soldiers started burning the buildings at one o'clock on the 13th. The ablebodied men were away with their militia units, and this made it difficult for those left behind to cope with the situation. The whole town was soon a mass of flames, the heat and flying brands igniting the furniture piled in the streets so all was lost. The Americans beat a hasty retreat to the river as British forces came rushing up to the town.

Some compensation was paid for the destroyed property later. It is an interesting speculation as to whether Niagara would have continued as an important commercial centre if not for the double blow of the removal of the capital and the destruction of its building stock.

Muddy Little York

Born long after the cities of Quebec and the Maritimes, Toronto went through the same early growing pains in fire protection half a century later. But it had some advantage from studying the developments that had taken place in other parts of Canada, the United States and Europe. Nevertheless, muddy little York as Montrealers called it, repeated much the same sort of faltering steps.

Governor John Simcoe named and started to lay out the Town of York on the bay where the French had their Fort Rouille and Fort Toronto, on August 27, 1793. Only after a few years, when it became definite that York was to be the capital of Upper Canada, did it begin to grow with any speed. The first fire on record in this outpost of civilization between Kingston and Niagara was noted by Mrs. Simcoe in her diary, when she described the burning of the cottage of Mr. Scadding on January 31, 1794. That is, if we do not count the burning of Fort Toronto by the French troops when they withdrew in 1756.

Some recent accounts state that the first step to fire protection was a bucket law passed in 1820. However, things began before that. According to the Upper Canada *Gazette*, the first bucket regulation was introduced by the legislature in 1800. The Act to Guard Against Accidents by Fire required each householder to maintain in his home two two-gallon buckets made of leather, wood or canvas, each carrying his name. Also, each house was required to have two ladders, one to the eaves and the other up the roof to the peak by the chimney. The ladders gave quick access to fires in the chimney or in the wood shingles of the roof, usually set by sparks from the chimney.

Furthermore, the Upper Canada *Gazette* on December 18, 1802, had this to say; "His Excellency the Lieutenant Governor has been pleased to give to the inhabitants of this town the use of a fire engine, for which signal mark of his parental concern for the safety and welfare of the King's subjects, all ranks and description of people seem highly grateful. As a very small token of their sense of this indulgence, a subscription was cheerfully set on foot, which immediately filled, for the erection of a proper building in the town for the preservation of the engine, and such measures will be immediately adopted as are best calculated to procure the most easy access to the engine at all hours of the day and night."

Evidently the citizens did not immediately become proficient at fighting fires with the engine as "A week later, Mr. James Playter's house on Yonge Street was entirely consumed by fire together with every article in it." No fire brigade was organized. The settlement was small and evidently everyone was expected to turn out and help. In July of the same year, the Grand Jury asked that all cabinetmakers, joiners and those who make shavings in their trades be obliged to burn or otherwise destroy same on Wednesdays and Saturdays at the hour of sunset.

Alexander Wood was an unofficial representative in York for Alexander Auldjo of Montreal, the first agent of the Phoenix Fire Office for Canada. He wrote Auldjo in 1805 complaining of the increases in fire insurance rates, explaining that York was made up of well-separated houses, unlike the more crowded Detroit, which had burned recently. Further, he noted, the fences joining the buildings could be torn away in five minutes to stop them from communicating fire between the houses. On top of this, by order of the Sessions, every householder had provided ladders and buckets. He offered to send a plan of the town if the company would pay for it.

There seemed to be numerous fires in this period, but because the houses were well scattered, there were no conflagrations. There was a continual threat from forest and grass fires that raged around the perimeter of the town in dry periods, threatening it with total extinction. Wood wrote to a friend the following May describing fires in the woods and fields getting into the fences, haystacks and woodpiles. They were fought by soldiers and citizens, who were always willing to help. There was no mention of firemen or a fire engine.

An inventory of the buildings in York in 1809 listed 55 one-storey frame houses, 14 log cabins, 11 one-storey and 27 two-storey squared timber houses. The first brick house came the next year. The streets then were rudimentary trails with numerous tree stumps still in them. There was a high constable appointed at £10 per annum and unpaid constables appointed from time to time by the magistrates. This was an unpopular duty, but refusal to take one's turn was punishable by law.

The War of 1812 came with a seaborne invasion, the Americans successfully landing on the beaches west of York to capture the town and occupy it along with Fort York. The powder magazine blew up in the battle, killing both American and British soldiers. The Americans are believed to have burned down the one-storey parliament house, although there is some doubt who did it, on April 8th, 1813. They also burned the barracks, shipyard and a few other buildings before departing.

When they left, the invaders took with them, as part of the booty of war, a fire engine,likely the town's first and, until then, only engine. It has been surmised that this may have been a fire engine of the army garrison, but this seems improbable as there was no mention of such an engine existing in connection with previous town fires.

It was the Royal Navy, of all entities, that came to the rescue at the fire which threatened the house of Dr. Strachan in February 1817. Captain Otty and the crew of the brig *Charwell* saved the house while the lieutenant governor directed operations to remove the furniture. This would seem to confirm that there was no fire company yet, nor any effective means of marshaling citizens at large to fight fires.

A new bucket law, or bucket brigade law, as it might better be called, was imposed in 1820. The householder was now to keep two or three buckets hanging outside his house, increased in size to three gallons and to be visible near the entrance. All citizens were to turn out on an alarm of fire. This was to be spread by means of "shouting, first solo and then in concert", and ringing the bell of St. James' Church, the only bell in town, all until a goodly crowd had gathered.

Additional means for providing fire protection were being taken from the evidence of a meeting of a Legislative Assembly committee. The Upper Canada *Gazette* reported February 4th, 1822, on the committee questioning Stephen Heward, Clerk of the Peace, on the disposal of over £300 in accumulated police funds. Heward said that he had bought a fire engine, among other things. A statute in 1817, providing for the establishment of police in York, Sandwich and Amherstburg, gave the local magistrates the right to raise funds for purchase of fire engines, buckets and other equipment up to £100 a year.

On further questioning, Heward said, "By report of competent judges, the engine is in good order, with the exception of a trifling breach in the supply hose which can be easily repaired." This engine was ordered from England in 1818 at a cost of £150. Heward also advised the committee that there was no fire company, as it was not possible to compel the attendance of those who had previously volunteered. The real problem was that the town could not promise the firemen the customary exemptions from parish duties such as compulsory road work, as was required of all citizens. It was the same difficulty Saint John had encountered 50 years before.

The various fire regulations were consolidated, revised and printed under the direction of Heward in 1823. Now there was an inspector of chimneys, who directed their cleaning every four weeks. Six fire wardens were appointed, two per ward, to command and enforce with the aid of the constables, the aid and assistance of all male inhabitants of the town from 16 to 60 years, to preserve order, regularity and dispatch in the operation of the bucket brigade lines. They were empowered to appoint guards for furniture and goods to preserve them from injury and pillage. Their badge was a white handkerchief worn as an armband.

(V) The Insurance and Military Connections

Origins of Firefighting Companies

Fire insurance from its beginnings was a valuable tool for the property owner to protect himself from calamitous loss by fire. As its main purpose, fire insurance eliminated the shock of a total loss of building and contents, sharing out the loss burden among thousands of fellow premium payers, with cash compensation to the unfortunate few.

But the fire insurance companies were also involved in other ways, through promotion of firefighting organizations from earliest times, whether it was by donating equipment or by exhorting the local authorities to improve their protective measures. The pioneer companies in England soon discovered that, with no public fire departments in existence, they took heavy losses and that they would have to take some initiatives in firefighting to protect their risks.

Hand-powered fire pumping engines were becoming a practical device at that time, the latter seventeenth century. The British insurance companies organized their own fire engine crews, employing the Thames watermen as firefighters. They then developed a system of identification for the buildings they insured and wished to protect. Plaques carrying the mark or coat of arms of the company were mounted on the face of each building. Firemarks were used in Britain and Europe for about a century before losing their original function. They continued in use largely as an advertising device, and spread worldwide, remaining in use in the Orient until World War II. Used in the older cities of the United States, firemarks are said to have advertised the fact that the building was insured and that the volunteer firemen would be rewarded if they performed well.

Elaborately equipped and uniformed paid fire companies were the style when the fire insurance companies organized the firefighting in Britain. Horses were enlisted to transport pumps to fires much earlier than in North America.

The British insurance company fire brigade became, in addition to a help in reducing the company's losses, an additional benefit to the policyholder by extinguishing his fire. Prominent companies maintaining their own brigades in London included the Royal, Guardian, Sun, Alliance and Phoenix. The companies fostered fire protection in other British cities by organizing similar brigades or by donating fire engines for the local governments or citizens to use.

Eventually, in 1833, the insurance companies combined their fire brigades into a single London Fire Brigade Establishment under the legendary Chief Braidwood, who pioneered much in the art of firefighting. This organization served in lieu of a municipal fire brigade for London until 1866 when the first publicly owned fire service was formed for London. Some British towns had insurance company-operated fire brigades until as late as the 1920s.

The development of such firefighting organizations as the British insurance brigades and municipal brigades in European countries like Germany and Holland represented a rebirth of organized firefighting after the Dark Ages. Imperial Rome had a firefighting system in the form of a night watch that fought fires in the city. It was known as the *Cohortes Vigilum*. Rome was not all monumental buildings, having masses of crowded wooden buildings housing the poorer classes. The vigiles died with the Empire, and European cities remained for centuries at the mercy of fires, except for the passive defense of their massive masonry construction.

The origin of firefighting organizations in North America was quite different from that of Britain, though the concept of a fire-engine-equipped crew undoubtedly came from there. It took a little time for the idea to take form in the youthful cities of British North America, though Boston was quick to act with its paid municipal firefighters in 1679 and volunteer companies 37 years later.

Most of the fire companies started in the pioneer communities of North America seem to have been derived much more from the spirit of community self-help, a local initiative reflecting directly the expressed needs and desires of the citizens involved.

In the same vein the first North American fire insurance companies were local mutuals, the first lasting one being the Philadelphia Contributionship, in whose organization Benjamin Franklin had a prominent role. Such homegrown efforts were necessary in the mid-eighteenth century, as the established British companies showed no inclination to provide insurance in the colonies.

Interestingly, it was the associations founded for mutual assistance in firefighting and salvage that spawned the first American insurance companies, quite the reverse of what had happened in England. Franklin was a prominent leader in the pioneer volunteer firefighting company, the Union of Philadelphia, in 1736. Within 14 years there were six such fire companies in that city.

Firemarks placed on buildings by insurance companies and mutual protection associations in the United States were not part of the Canadian scene. This well known Hand in Hand firemark symbolizes the mutual help theme of the early protection clubs and associations.

Boston is accepted as having had the first fire department in the United States, in 1679, this being a group of men paid to man a hand fire engine imported from England. Mutual fire societies began in Boston in 1718 and in general helped to foster the North American volunteer fire company. These mutuals were later known as protection clubs or companies, and still later as salvage companies in Canada. In Boston they had the purpose of saving the contents of burning or threatened buildings while the paid firefighters fought the fire under sponsorship of the insurance companies.

Benjamin Franklin carried these ideas with him when he moved from Boston to Philadelphia. They were brought to Canada in significant part by the Loyalists, it may be assumed, although communication among New England, the Maritimes and pre-Confederation Canada was fairly good before their arrival.

Self-help in protection against fire on a volunteer, neighbour-help-neighbour basis was in the style of the North American pioneer. From it the volunteer fire services of Canada and the United States have descended and remained important even today. However, the need for fire insurance on a substantial scale that would allow owners of large value properties to be protected against calamitous loss was evident.

Fire Insurance Arrives In Canada

The British insurance companies did not hang back indefinitely. One company that had pioneered in foreign markets was the Phoenix Fire Office, direct successor to the first fire insurance company, "The Fire Office," founded in 1667, immediately after the Great Fire of London. The Phoenix was persuaded to accept a risk in St. John's, Newfoundland, in 1782, and one in Halifax in 1784. Requests for insurance elsewhere resulted in the Phoenix appointing the first agents in Canada, Alexander Auldjo of Montreal becoming their representative for Upper and Lower Canada in 1804. The next year Forsyth, Smith and Company of Halifax were made agents for Nova Scotia, and Messrs. Marmaduke, Hart and Company agents for Newfoundland.

Along with the issuance of fire insurance policies came the practice of encouraging local fire protection through donation of fire engines and equipment to the town authorities. Other major companies followed the Phoenix to Canada and they also donated or contributed towards fire engines and related equipment, as did Canadian companies which appeared on the scene, these including the Halifax, the Provincial and the Montreal Mutual. In Toronto the insurance companies' generosity turned out to be a loan. The Provincial and the British America reclaimed their engines in the 1850s after they had been retired from active use by the Toronto Fire Brigade.

While British insurance companies were doing business soon after 1800 and sometimes offered help to Canadian companies (the Alliance had several engine companies carrying its name) there is no indication that any of them actually operated a firefighting company in Canada. The one known case of an insurance fire engine company was that of Quebec Fire Insurance Company. This was founded in 1827 and operated an engine manned by employees as firemen, placing firemarks on buildings it insured in the City of Quebec.

The Quebec Fire Insurance Company was incorporated in 1816, the event being marked by a Masonic ceremony and parade when the office was opened for business two years later. The Masonic connection is intriguing. Perhaps the Masons had a direct hand in its organization.

Homegrown Canadian insurance companies also rose to do business. Munson Jarvis of Saint John, a well-to-do shipowner, was, in 1801, carrying on an insurance business, first in marine coverage and then in fire. His descendants carried on, but switched to being agents for British companies. The Fire Insurance Association of Halifax was founded in 1819. This organization performed a function as advocate for improvements in fire protection. It was the progenitor of the Halifax Insurance Company in Canada. This company gave a fire engine to the Union Engine Company of Halifax in 1827. About this time, The Mutual Insurance Company of Montreal was also established.

LONDON,
1804.

PROVINCES
OF
UPPER AND LOWER
CANADA,
NORTH AMERICA.

PROPOSALS

FROM THE

PHŒNIX COMPANY OF LONDON

FOR INSURING

HOUSES, BUILDINGS, STORES, SHIPS IN HARBOUR, GOODS, WARES, AND MERCHANDIZE, FROM LOSS OR DAMAGE BY FIRE.

INSURANCE from Loſs or Damage by FIRE hath been found a Meaſure of great Importance to the Happineſs of Families, and has given additional Security to Commercial Tranſactions. The diſtinguiſhed Approbation with which the Public throughout GREAT-BRITAIN received the improved Syſtem of this Office, occaſioned frequent Application for the Inſurance of Property in the principal Cities and Towns in *Europe* and *America*, and induced the Company many Years ſince to extend their Plan to the WESTERN CONTINENT. The PHŒNIX COMPANY can refer to the unerring Teſt of Experience to aſcertain their Character; and, gratified by the unbounded Confidence they have obtained upon the Continent of *America*, and for the Purpoſe of facilitating to the Inhabitants the Means of effecting Inſurance on their Property, the DIRECTORS have eſtabliſhed an AGENT in CANADA, and have arranged Rates of Premium upon the various Deſcriptions of Property, which, they truſt, will be found moderate, juſt, and equitable. The Promptitude with which this Office has adjuſted the Claims of Sufferers, and the Solidity of its Funds, are ſo well known, that it is unneceſſary to offer more on the Subject, than to refer thoſe who deſire Information to the Merchants of *London* with whom they correſpond.

RATES of ANNUAL PREMIUMS to be paid for ASSURANCE againſt FIRE.

	No. I.	No. II.	No. III.
	Hazards of the Firſt Claſs, viz.	*Hazards of the Second Claſs, viz.*	*Hazards of the Third Claſs, viz.*
	Brick or Stone Buildings, covered with Tiles, Slate, or Metal.	Buildings having the Four Walls entirely of Brick or Stone, carried through the Roof, and covered with Boards or Shingles.	Timber or Slight Buildings covered with Shingles, or Boards;
	Furniture or Merchandize not hazardous, contained in ſuch Buildings.	OR, Buildings having the *Side* Walls of Brick or Stone, carried through the Roof, and the Roof covered with Tiles, Slate, or Metal.	ALSO,
		Furniture or Merchandize not hazardous contained in ſuch Buildings.	Hazardous Goods, viz. Pitch, Tar, Turpentine, Salt-Petre, Flax, Hemp, Oils, and Tallow, in Buildings of the Second Claſs.
		Hazardous Goods, viz. Pitch, Tar, Turpentine, Salt-Petre, Flax, Hemp, Oils, and Tallow, in Brick or Stone Buildings.	
For Sums not exceeding	English Pounds Sterling. 1000 *in one Riſk* 6s. *per Cent. per Annum.* 3000 *ditto* .....7s. *per Cent. per Annum.* 5000 *ditto* .....8s. *per Cent. per Annum.*	English Pounds Sterling. 1000 *in one Riſk* 9s. *per Cent. per Annum.* 3000 *ditto* .....10s. *per Cent. per Annum.* 5000 *ditto* .....11s. *per Cent. per Annum.*	English Pounds Sterling. 1000 *in one Riſk* 15s. *per Cent. per Annum.* 3000 *ditto* .....16s. *per Cent. per Annum.* 5000 *ditto* .....17s. *per Cent. per Annum.*

☞ Tallow-Melters, Soap-Makers, Brewers, Vinegar and Sweet Makers, Hemp and Flax Dreſſers, Printing-Houſes, Coopers, Carpenters, Cabinet-Makers, Coach-Makers, Malt-Houſes, Bakers, Ship-Chandlers, Boat-Builders, Rope-Makers, Sugar-Refiners, Diſtillers, Chemiſts, Varniſh-Makers, Turpentine-Works, Theatres, and all Mills and Machinery, are deemed extra-hazardous, and the Rate upon ſuch Riſks will be proportionably increaſed.

CONDITIONS OF INSURANCE.

I. PERSONS deſirous to make Inſurance on BUILDINGS, are to deliver in to the Agent the following Particulars, viz.—Of what Materials the Walls and Roof of each Building are conſtructed—Whether the ſame are occupied as private Dwellings, or how otherwiſe—Where ſituated—Alſo, the Name or Names of the preſent Occupiers.

Each Building muſt be ſeparately valued, and a ſpecified Sum inſured thereon—and in like Manner a ſeparate Sum inſured on the Property contained therein.

All Manufactories which contain Furnaces, Kilns, Stoves, Coakels, Ovens, or otherwiſe uſe Fire-Heat, are chargeable at additional Rates.

In the Inſurance of GOODS, WARES, or MERCHANDIZE, the Building or Place in which the ſame are depoſited, is to be deſcribed; alſo, whether ſuch Goods are of the Kinds denominated Hazardous, and whether any Manufactory is carried on in the Premiſes. And if any Perſon or Perſons ſhall inſure his or their Buildings or Goods, and ſhall cauſe the ſame to be deſcribed in the Policy otherwiſe than as they really are, ſo as the ſame be charged at a lower Premium than is herein propoſed, ſuch Inſurance ſhall be of no Force.

II. Goods held in Truſt, or on Commiſſion, are to be inſured as ſuch, otherwiſe the Policy will not extend to cover ſuch Property.

III. No Loſs or Damage to be paid on Fire happening by any Invaſion, Foreign Enemy, Civil Commotion, Riot, or any military or uſurped Power whatever; nor for Damage done by Fire occaſioned by Earthquakes or Hurricanes: but this Company will make good Loſſes on Property burnt by Lightning.

IV. Books of Accounts, Written Securities, Bills, Bonds, Tallies, and Ready Money, cannot be inſured.

V. Jewels, Plate, Medals, or other Curioſities, Paintings, and Sculptures, are not included in any Inſurance, unleſs ſuch Articles are ſpecified in the Policy.

VI. Perſons inſuring Property at this Office, muſt give Notice of any other Inſurance made elſewhere on their Behalf on the ſame, and cauſe ſuch other Inſurance to be endorſed on their Policies; in which Caſe each Office ſhall be liable to the Payment only of a rateable Proportion of any Loſs or Damage which may be ſuſtained; and unleſs ſuch Notice is given, the Inſured will not be entitled to recover in Caſe of Loſs.

VII. No Order for Inſurance will be of any Force, unleſs the Premium is paid to the Agent, or unleſs a Sum has been advanced, and the Agent has delivered his Receipt on Account of the Office; and all Perſons deſirous to continue their Inſurances, muſt make their future Payments Annually within fifteen Days after the Day limited by their reſpective Policies, or the ſame will be void.

VIII. All Perſons aſſured by this Company, ſuſtaining any Loſs or Damage by Fire, are forthwith to give Notice to the Company's Agent, and as ſoon as poſſible after, to deliver in as particular an Account of their Loſs or Damage, ſigned with their own Hands, as the Nature of the Caſe will admit of, and make Proof of the ſame by their Oath or Affirmation, and by their Books of Accounts or other proper Vouchers, as ſhall be reaſonably required; and ſhall procure a Certificate under the Hands of a Magiſtrate or Sworn Notary of the City or Diſtrict in which the Fire happened, not concerned in ſuch Loſs, importing that they are acquainted with the Character and Circumſtances of the Perſon or Perſons inſured, and do know, or verily believe, that he, ſhe, or they, really, and by Misfortune, without any Kind of Fraud or evil Practice, have ſuſtained by ſuch Fire, Loſs and Damage to the Amount therein mentioned; and until ſuch Affidavit and Certificates are produced, the Loſs Money ſhall not be payable; alſo, if there appears any Fraud or falſe Swearing, the Claimant ſhall forfeit his Claim to Reſtitution, or Payment, by Virtue of his Policy.

IX. In Caſe any Difference or Diſpute ſhall ariſe between the Aſſured and the Company, touching any Loſs or Damage, ſuch Difference may be ſubmitted to the Judgment and Determination of Arbitrators indifferently choſen, whoſe Award in Writing ſhall be concluſive and binding to all Parties. And when any Loſs or Damage ſhall have been duly proved, the Inſured ſhall receive Satisfaction, to the full Amount thereof, without Allowance of any Diſcount, Fees, or other Deduction whatever.

X. Perſons chuſing to inſure for Seven Years, will be charged for Six Years only; alſo, for a leſs Number of Years than Seven, will be allowed a reaſonable Diſcount.

⁂ Perſons deſirous of effecting Inſurances are requeſted to make Application to ALEXANDER AULDJO, Eſq. of *Montreal*, the Company's Agent for *Canada*, who is fully authoriſed to grant Policies on Behalf of the Company.

The Phoenix Assurance Company was descended from the first British fire insurance company and was first to take on business in Canada as shown by this announcement of rates.

To see to its interests when venturing to do business in the wooden cities in Canada, the Phoenix sent out inspectors to evaluate and report back on conditions, as well as counsel their resident agents. The longhand reports of Jenkin Jones and John Broomfield in 1808 and 1845 — preserved in the company's files still — shed considerable illumination on the hazards, the protection and the fires in Canadian cities in those times. They provide us with an excellent feeling for conditions and attitudes relative to the fire question

Jenkin Jones arrived in North America via New York and proceeded to Montreal. On August 7th, 1808, he wrote home to his London office giving his impressions. He noted the wood-frame buildings in the villages and countryside enroute were similar to Sweden's. The difference was that in New York State, which he travelled through, the houses were well separated. He concluded that the Americans were much more careful to prevent fires and more expert at extinguishing them. Few places in New York were without fire societies, he said. (Jones would hardly consider the Americans or Canadians very careful about fire prevention if he were to compare them with Sweden today.)

Arrived in Montreal, the inspector for the Phoenix decided that Canadian towns had the advantage of snow cover on the roofs a good part of the winter half of the year, when candles and heating fires are most in use. This should help reduce the danger of fires spreading by way of wood shingle roofs. He noted the requirement for ladders to reach the chimneys.

Jones observed that both Montreal and Quebec were well supplied with fire engines at that time. His company had, in fact, presented a fire engine to the Society for the Prevention of Fires, in Quebec, in 1790, and this was acknowledged in the Quebec *Gazette* of July 29. Montreal received a fifth-size engine, donated in 1804 and delivered the following year. Trois Rivieres also received a Phoenix engine about this time and named it "La Protection."

Reaching Montreal on August 6th, 1808, the inspector for the Phoenix was favourably impressed. Writing to his home office, he described the principal city of Canada thus, "The town is an old-fashioned French city, ill laid out, the houses almost all of very solid masonry, and the only important risk is from a certain proportion of them being covered with boards. The great number of fire proof buildings, every third house almost being of that construction, certainly prevents every danger of a fire spreading to any extent. I cannot imagine it possible for more than three or four houses to be destroyed by one accident.

"You are aware that in the covering of houses in this country they never employ tiles or slates, which are not thought to stand the climate as well as tin or sheet iron.
The number of houses, churches and convents covered with tin and protected with with heavy shutters and iron doors give the town a most gloomy appearance, but confer full security against those dangerous fires which were anciently experienced in Montreal."

"The suburbs as you might imagine are generally built of timber, tho' there are both in the St. Lawrence and Quebec Suburbs, some very substantial buildings. We have a mere trifle insured there, the houses being very small, of course of little value and generally inhabited by the poorer classes who cannot afford to insure them."

Jones was referring to the compact central part of the city, the commercial core, when praising the solid fire-resistant construction. This was primarily the crowded, walled city built by the French long before and known today as Old Montreal. The St. Lawrence and Quebec suburbs were the areas lying to the north and east of the old city. When Jones visited, Montreal had not yet been struck by the waves of immigrants from across the Atlantic who would so greatly expand its size through the middle of the nineteenth century.

Proceeding down the St. Lawrence River to the City of Quebec, Jones was obliged to report, "Of Quebec, I can by no means report in the same favourable terms, and we must have acted sometime since in a strange error in charging it at lower premiums than Montreal." He explained how Upper Town was located on a bluff about 200 feet above the Lower Town and thus had little connection with it except that it was very vulnerable to sparks and flaming brands from a fire in the higher section of the city.

While the buildings were almost exclusively of solid masonry construction, the roofs were of wood. The lowest fire risk was in the area of upper-class homes on the uphill side of Upper Town, towards the Citadel. Here there was ample spacing and tin-covered roofs, though the iron shutters and doors of Montreal were lacking. The other side of rue de la Fabrique on the other hand, was more hazardous, "inasmuch as it is closer built and less respectably inhabited. There are in it hardly any tin roofs and the streets are too narrow." With the constant presence of sentries and the permanent military force always at hand, Jones concluded that there was not too serious a risk of any fire becoming "formidable" in Upper Town.

Of Lower Town, Jones had this to report to the officials of the insurance company back in London: "The Lower Town of Quebec, or at least the greater part of it, beginning with Sault au Matelot Street and extending to the King's Wharf at the end of Champlain Street, I consider a most hazardous body of property. The streets which compose it are all extremely narrow. St. Peter Street, the best and broadest, being not more than 30 feet in width. The houses are crowded upon each other and with only two exceptions are covered with boards."

"Besides this, the Wharfs at the back of St. Peters Street are piled with staves, boards, etc. There are on them some large wooden stores in which are frequently deposited quantities of spirits. In fact St. Peters Street is the Wapping of Quebec and the whole district I

The idea of volunteer salvage or protection companies persisted in a few cities and some still exist. This is No. 2 Salvage Corps and Fire Police at Saint John, N.B., about 1900.

have described is so huddled together that there is no foreseeing where, under a concurrence of unfortunate circumstances, a fire might terminate."

Jones notes that flaming brands had already set roofs on fire in Lower Town from fires in Upper Town. He felt that this district, where so much valuable property was concentrated, could be safeguarded without too great an expenditure simply by covering the roofs in tin and providing iron shutters for the doors and windows.

Jones took drastic action and advised London, "In consequence of the danger to which I consider the main body of the Lower Town exposed and being persuaded it will be the wish of the Directors to be guarded against every chance of serious misfortune, I have desired Mr. Auldjo to decline all insurances in it, whether by new policies or renewals, and I trust the Board will approve this proceeding." The Phoenix was insuring some 60 buildings in the district to the tune of £60,000.

As far as Ontario was concerned in 1808, Jenkin Jones said, "I have had no opportunity of visiting Upper Canada, to have made my route through which would have consumed more time and incurred more expense than I felt myself authorized to bestow upon it. I have consequently not seen Kingston, nor York, but we have not in either anything considerable."

The British Military Garrison's Role

In the days when British Imperial troops were stationed in fortified and strategic cities, and local volunteer fire companies were primitive or nonexistent, the garrison normally turned out to help fight fires in the towns. References are frequent to the troops and the garrison engine being "quickly on the scene," along with or in lieu of municipal fire companies. It seems certain that the military demonstrated to the citizens and their councils that with organization and engines fires could be fought successfully, spurring the step from purely salvage-oriented fire protection societies to effective extinguishment. The difference between dashes of water from buckets and a steady, directed stream was more significant than it may appear.

Halifax was fortunate in having both naval dockyard and army garrison fire engines turning out both before and after the first town hand-engine companies were formed. The garrison buglers provided a clear and efficient alarm signal heard both in barracks and in town.

In 1817, apparently in response to an inquiry from the city fire wards regarding military assistance, the deputy adjutant general published new general orders covering army response to fires. The orders required all guards and *picquets* to be under arms, all officers and men not on duty to return, not under arms, to barracks for orders to march to the fire. They were "to form ranks and assist in passing water or otherwise as the fire wards may desire." The military fire engines were to turn out.

"The Captain of the Day and Field Officer will concert with the fire wards and magistrates, employ the *picquets* in the most advantageous manner for the security and the protection of property." The Town Major was to patrol the contiguous streets and "take up any straggling soldiers who may be found out of the stations allotted to them." A supply of one hundred buckets was placed near the guard rooms in the north and south barracks and were to be used only in the event of fire or when the engines were taken out.

As the civilian volunteer companies developed in numbers and organization, and fires became more frequent as the city grew, it became more onerous and less reasonable for the army and navy to run to town fires. The press reported on October 26, 1837, that while the town engines and those of the dockyard and ordnance responded promptly, the goodwill of the military was being rather tried by having to furnish the manpower when there was a poor turnout of volunteers. In addition, the soldiers found themselves working under the orders of the civilian fire wardens. The soldiers were reported on this occasion as berating one of the wardens, demanding, "Why don't you get your own men!"

Nevertheless, as late as 1857, the Halifax garrison saved the day in a 22-hour fire on Prince Street that burned St. Matthew's Church. It was 4 p.m. on New Year's Day, with many of the firemen out of town for the holiday. Water pressure in the hydrants was weak and there was much pilfering of goods. Help came from another source ten years after this at the Moir's Steam Bakery and Grist Mill fire. Seamen from a visiting squadron of French warships came ashore to help, bringing their portable fire engines.

In Saint John the garrison ordnance engine was looked down upon by the town firemen for its small size. It was said that it did not have steering front wheels and had to be lifted to turn a corner, and when in a hurry the soldiers would pick it up and carry it over rough parts of the streets. It responded from the south end barracks to about half of the city, that below Princess Street.

Kingston was the principal naval and military depot of the upper province, to become Ontario. In the 1840s, the town had several fire engines and men to man them, but according to the report by J.J. Broomfield of the Phoenix Insurance Office in 1845, "Chief dependence is upon the military who turn out instantly and work the engines very effectively."

The Black Watch Regiment, in garrison at Sydney, fought that town's first major fire, November 28, 1851. The outbreak occurred in a vacant building and spread to destroy three others. The Highlanders fought the fire with their engine impressively enough that the town thereupon bought a secondhand engine in Halifax with money raised by subscription. This engine proved unsuitable and, as in the case of Saint John, the town could not afford another. However, the garrison shortly turned over theirs to the town and it served for many years. Water was obtained from wells and the markethouse bell served as the alarm.

In general, the military had a strong presence in the pioneer cities that were of sufficient strategic importance to have a garrison. This is certainly reflected in the response of troops to maintain order and to fight fires, both in the French and British regimes, until the withdrawal of the Imperial troops and the growth of civil fire and police forces.

The major cities of the late 1700s were beginning to form a worthwhile measure of defense against fires. The bigger they grew, the greater the congestion of buildings and goods — with a matching dependence by the increasingly complex community upon the goods, services and amenities provided. Thus, as compared to the small towns and villages, the cities had relatively more at stake and greater perceived need, as well as greater resources to draw on. Similar factors apply in the modern fire protection picture.

In this situation, smaller places did not always rush to emulate their bigger brothers, but lagged somewhat, suggesting that there was a considerable tolerance of the periodic total destruction of individual buildings.

Newer, smaller towns like Fredericton, Yarmouth, St. John's, Trois Rivieres, Kingston, Toronto, Hamilton and London commenced working their way through the stages of fire wards, chimney and bucket laws, protection clubs and then fire engines a decade or more after 1800. In the meantime, the fire insurance companies had arrived on the scene to provide another kind of protection against fire. They also stimulated the growth of public protection measures in their own as well as the citizens' interests.

2

FIRE CLUBS TO FIRE DEPARTMENTS (1800-1840)
(VI) The Young Metropolis

A Conglomeration of Companies

As a city achieved a measure of firefighting capability, so that fires were actually being extinguished, the citizenry frequently seem to have relaxed vigilance, content that the necessary had been done. Complacency would last as long as reasonable luck held, and it would take a vicious slap or two from demon fire to drive home the fact that they had neglected to keep their protection in pace with the growing hazards.

Ignorance, entrenched interests of the firemen or the merchants, and the acceptance of adversity as a feature of existence were probable elements in tolerance of weak fire defenses in those times. It seems clear, in Halifax for instance, that good organization even then could have produced a superior fire record, with simple equipment resources. All this has had application down through the centuries in some degree. An unchanging human foible has left a pattern of fits and starts in the progress of fire protection.

From the example of the bigger, more sophisticated American cities, the military and the insurance companies, citizens and municipal governments took up the initiative to organize fire clubs and fire engine companies. Eventually officialdom established cohesive fire departments as an arm of municipal government. This was a perceived need, the service being technically an option and not obligatory upon the municipality even today. Examples of the steps taken by the principal cities illustrate the process. And still most of them were racked periodically by great fires.

As the biggest city, Montreal led the way in size and sophistication of municipal fire protection organizations. Snatches of its story in the early part of the eighteenth century show how it expanded far outside the stone-walled French town in sprawling wood-frame districts. Even the insurance inspectors were a little complacent.

As Montreal grew into the metropolis of Canada, from 1800 onward, spreading its streets far beyond the original fortified gravel ridge by the St. Lawrence River, the number of fire engine companies naturally grew also. In 1804, when the Phoenix Fire Office appointed as their agents for Canada Maitland, Garden and Auldjo, they arranged that Alexander Auldjo present to the governing magistrates of the city a little three-wheeled handtub fire engine. This was a single, 5-inch-diameter cylinder English model of very modest type. It is preserved still in Montreal. The magistrates turned it over to the fire committee with the request that they build an engine house for it. The engine was kept in the churchyard of Notre Dame until a small shed was erected to house it nearby.

Another, much larger, Tilley engine, also English, carried the name Alliance, having been donated by the insurance company of that name in 1826, the company establishing itself in Canada not long after the Phoenix.

By 1819 there were five fire engine stations: at the central square of Place D'Armes, at the fish market on St. Paul Street, at the corner of Notre Dame and St. Jean Baptiste, at the Recollet Church on Notre Dame Street, and at the Nunnery Gate on St. Joseph Street. These represented the five wards and each had its fire company attached. This was the area known as Old Montreal, a crowded history-steeped district in which no fire station operates today.

This little three-wheeler was presented to the city of Montreal in 1804. Water was poured into the body, or tub, and pumped through the single length of rivetted leather hose.

The housing of the five engines must have been rudimentary where it existed, as "the first real fire station" was erected near Notre Dame Church at Place d'Armes by the Association of Volunteers in 1824, under Antoine LePage. This association seems to have represented a step towards bringing the fire companies together as a single department. The population of Montreal was now about 10,000. Still, all the water for the fire engines was brought by bucket from the public well in Place d'Armes, from the St. Lawrence River or the Little River, running along Craig Street on the back side of town.

A picture of what the old firefighting organization was like in the 1830s, before the city took an active hand, comes from the reminiscences recorded by Alfred Perry, easily the most prominent figure in the Canadian fire protection world during the nineteenth century. This remarkable man first came to public attention as a firefighting hero at a Montreal fire that threatened the old Bonsecours Church in 1832, or the following year, when Perry was aged 12 or 13. He died in office as Fire Commissioner of Montreal in 1900.

Perry wrote disparagingly of the state of the fire department prior to 1842: "At the time the old apparatus was transferred to city control there was no fire station with a stove. Hose was piled up to thaw when a sunny day came. Fire engines were left in the sheds that served as stations and were permitted to freeze after use. When there was a fire it was necessary to invade a nearby home for hot water to thaw the pump valves, a delay of perhaps half an hour. When one company invested 13 shillings in a tin stove for their engine house, the authorities refused to pay for fuel for it."

The five fire engines were of rather primitive type, according to Perry. There was the big heavy Tilley engine from England. Weighing over a ton, it pumped only 40 to 60 gallons per minute through a five-eighths-inch-diameter nozzle. Two others were of gooseneck style and probably the city's original engines, capable of throwing a stream of water from the nozzle on top of the engine, and thus incapable of fighting fires well inside a building. The little Phoenix at least had a length of leather hose that allowed this.

There was at that time no hook-and-ladder company, but 25- to 30-foot ladders were located conveniently about town. There was apparently a property protecting company, though little has been recorded of it. It would have been equipped with bags so that members could easily load up removable valuables and take them away for safekeeping.

An old French ordinance typical of the day required every householder to keep two buckets available and to produce them when there was a fire in the neighbourhood. It was necessary to have one's name on the buckets, so as to be able to reclaim them from the common pile afterward. They were intended to be used in a bucket brigade chain to carry water to the fire. This was largely replaced when the system of commercial water carters became well established.

Before a successful piped waterworks system was installed in 1832, householders purchased their water by the bucketful from carters, who had a puncheon on a two-wheeled cart and trucked the water from the St. Lawrence River. Larger ones were pulled by horses, some small ones by dogs. Sleds were used in winter. There were between 50 and 100 water carters in business even after the first waterworks system was in use, perhaps largely serving the outlying areas. Still, when the alarm of fire was given, the water carters responded, spurred by a premium for the first load delivered. The slowness in getting water from the fireplug pits on the watermain caused the carters to remain an asset.

There was great rivalry among the companies in getting to the fire first, both for the honour and the five dollars prize for getting first water on the fire. Fred Perry and his brothers of the Union company seldom lost out in this contest at night — by the simple expedient of bribing the police, who spread the alarm to notify them first. This racket ended when the alarm by the police night patrol was abandoned for a system of having the firemen take turns on night watch duty in their stations. In winter these "sleepers" would frequently run out and seize any passing horse to haul their engine through the snowdrifts, as was their perogative.

Predecessor of the present Notre Dame Cathedral, this little church provided Montreal's early fire alarms by its tower bell. It was only necessary to rouse out the beadle to set the bell tolling for the volunteers.

- PAC-C12531

The first fire alarm device used in most cities and towns in North America was the church bell. In Montreal it was the bell in the original little church in Place d'Armes that was the predecessor of the present Notre Dame. Long after the original church was demolished, the tower remained, since it was many years following the construction of Notre Dame that its twin towers and great bells were installed, in their turn, to become part of the Montreal fire alarm system. The bells of the Anglican Cathedral and other churches eventually joined the chorus.

A City Fire Department

A more significant step in the organization of fire protection in Montreal came with the establishment of the Fire Society or, more properly, The Society for the Prevention of Accidents by Fire, under a provincial charter in 1839. This independent, self-governing entity was granted powers to enforce fire prevention regulations in such areas as chimney cleaning, smoking in hazardous locations and burning of refuse out of doors, and was to collect fines for infractions. Two previous provincial acts for fire prevention were partially suspended in this transfer of responsibility to an association comprised of 30 building proprietors under a chairman. The curious existence of this enforcement body may have been connected with the fact that the city's charter of 1833 had been suspended since the French Rebellion of 1837. It could only have been awkward for a group of businessmen to police a set of fire regulations.

The act also provided for appointment of up to 50 volunteer firemen to man the fire engine in each ward under the direction of a captain. These firemen were to be exempt from military duty except in an invasion and from road work or the contribution of money in lieu of such work. There was also an auxiliary company of men that may have been a protection company, to which there were some allusions around this time. This group covered the whole city in any case.

The regulatory fire prevention duties of the Fire Society proved too much for this kind of management. The city regained its charter in 1841 and took over responsibility for the fire companies and fire prevention. At this point continuous fire protection began under a chief appointed by and responsible to the city council. A bylaw of June third created a fire department under the direction of an inspector, a superintendent and a chief engineer, with one or more overseers of the sweeping of chimneys, as well as one captain and lieutenant for each fire company, to be appointed by and under direction of the fire committee of council. There were to be 20 to 25 men in each company, and these to be selected by the inspector. This last official has not been explained, but seems likely to have been a police official, as it was in line with British habit at the time to place firefighting under a duly constituted authority such as the police.

Montreal's first fire department chief engineer was Joseph R. Bronsdon, who was appointed March 10th, 1841, and is accepted as having been the first fire chief in the continuous line of succession down to this day. The changeover to a city-controlled fire department appears to have taken effect in 1842.

A daily diary kept by a businessman of Montreal, and later Quebec, Abraham Joseph, provides a record of several fires in Montreal with the old fire department. On 13 July 1835, Joseph wrote, "At halfpast ten o'clock the bell of the English Church gave the alarm of fire, which turned out to be in the stores of Messrs. G. and N. Dacond and the auction rooms of L.L. Pinsonaut with stores situated in the same building in St. Paul Street, near the old market. The different fire companies were in attendance soon after the alarm was given, but before they could be got into operation the house was nearly burnt. No water carters were in attendance with the exception of one belonging to the nuns, and perhaps a couple of others.

"An engine was placed on the brink of the wharf, which, with a hose, supplied one placed in the rear of the building in Capital Street. Another engine was placed in Messrs. McGill & Co.'s yard. The fire originated from the store of Dacond. The progress of the fire was such, that scarcely anything was saved, not even the books and papers. It is stated that the estimated loss of Messrs. Dacond exceeds £4,000, besides their books, in which were outstanding debts to the amount of £9,000.

"Insurance was effected at the Etna Office for only £1,500. Mr. Pinsonaut's loss was considerable and he had nothing insured. The house belonged to Wm. Cormack, Esq. and was insured at the Alliance office for £1,500."

Two weeks later, on the 21st, Joseph reported a "rekindle," always an embarrassment to a fire department. "Last night at seven o'clock the premises of Messrs. P.McGill & Co. were discovered to be again on fire. The flames from the adjoining house which was burned on the 13th appear to have been making their progress quietly and imperceptibly thro' the wall of Mr. McGill's warehouse & burst out in a room in the second storey. The immediate operations of the firemen & a supply from the cellar of Mr. Newbold soon extinguished the fire. Very little damage was done. " A couple of weeks later Mr. Joseph was to report the death of three workmen and four horses in the collapse of a wall from one of the burned buildings while clearing up the ruins.

Just two days later, a diary entry reports a fire breaking out in Ward's Foundry and, from the extreme dryness of the wood, it was completely enveloped in flames almost immediately and before the fire engines could arrive. On August sixth Joseph made the laconic entry, "At half past 3 a.m. a fire broke out in McGill Street at the entrance of Recollet Suburbs in a place occupied as a hat factory. It being built of wood, was immediately burnt to the ground — no insurance."

Another Look by the Phoenix

The Phoenix Fire Office of England, the pioneer and likely largest insurer against fire in Canada at the time, sent out another inspector in 1845 to visit their agents and report firsthand on conditions in North America. In 1808 Jenkin Jones had given a rare picture of our pioneer cities' fire hazards. His successor, John J. Broomfield, wrote a series of similar reports 37 years later that provide us with another set of rare appraisals of our major cities as fire risks.

Canada's largest city impressed the inspector for the Phoenix in 1845 thus: "Montreal is in a very thriving condition, her merchants are enterprising and active, both trade and commerce is being pursued with alacrity, and new buildings for the accommodation of the increasing number of inhabitants are being erected in all directions, but more especially in that part which lies to the westward of the St. Lawrence and St. Antoine Suburbs."

Although Montreal had spread far beyond the original walled town on the waterfront into new adjoining areas referred to as suburbs, that old congested area remained the downtown commercial heart on which Broomfield concentrated his attention. This district, as the insurance surveyor noted in a report dated October 4th, 1845, included the mercantile part of the city and the most elegant and substantial of the public buildings. "The houses in this district," he advised his company, "may be said to be wholly constructed of brick and stone, there being scarcely a timber building in it, although the outhouses or sheds for domestic purposes are almost invariably of timber construction."

Pride of performance in running with their "hand-tubs" led the volunteers to races, collisions and fights. Horses were long spurned, except in heavy winter snows.

Broomfield reported, as had Jones before him, that the buildings were well built, with good masonry walls dividing them, and no external woodwork, such as clapboarding, throughout the whole district. The roofs were nine-tenths covered in tin or sheet iron. The balance were unfortunately in wood shingles. The merchants' stores along narrow old St. Paul Street were fitted with iron window shutters and doors as protection against neighbours' fires, as were major warehouses elsewhere in the city. The practice of installing these devices had been discontinued on new construction, however, a case of evident complacency in the absence of recent conflagrations.

"Notre Dame Street," he said, "is occupied principally with the shops of the retail dealers, some of them handsomely fitted up and would bear comparison with the first class shops in London." Great St. James Street, the fourth of the main east-west thoroughfares running parallel to the St. Lawrence waterfront, was 70-feet wide and contained several handsome public buildings "in governmental architectural style." It was, in fact, on its way to becoming the great financial street of Canada for one hundred years. Most of the major wholesale warehouses were located along the waterfront on Commissioners Street.

Broomfield considered the only major concern to be the wooden back sheds, but observed that a new provincial law now required that these and any wooden fences in the old city be removed within two years, substituting only brick or stone structures with tin-clad roofs. He had some concern for the narrowness of some of the north-south side streets, but in retrospect was overly confident in believing the spread of fires from building to building was unlikely.

Perhaps Broomfield was blinded by the contrast with the suburbs, for as soon as one passed Fortification Lane, a street north of St. James formerly occupied by the city wall, there was a sea of wooden buildings with wood-shingled roofs fronting on Craig Street and streets beyond. This was to be a common condition in Canadian towns, a sudden deterioration of quality of construction immediately beyond the boundary in which building regulations were imposed. Many of the common walls between the masonry buildings were not as good as Broomfield thought, some were rubble-filled, without much mortar and probably of a flawed fire-resistance value. Great fires at the mid-century would prove his confidence misplaced.

The biggest single-risk account for the Phoenix in the old city was the stately Notre Dame Cathedral located on the south side of Place d'Armes square and insured for £10,000.

As for firefighting, Broomfield stated, "I am happy to find that Montreal is well provided with the means for extinguishing a fire. Throughout all the City District the water is laid down in the streets, at a sufficient depth so as to protect the pipes from frost, and conducted into iron standards at the corner of almost every street so that in case of fire a plentiful supply of water can be obtained instantly. And an act has been passed authorizing the Corporation to construct twelve large tanks for the supply of other parts of the city and which are now being proceeded with.

There is also a well organized body of firemen who are exceedingly active, and they have some good working engines, the prompt attendance of which is secured by a reward being given to the first that arrives."

Broomfield said he had an opportunity to see the Montreal firemen in action one Sunday. The fire broke out in a cooperage shop in the southwesterly St. Ann's suburb. When he arrived the cooperage, a wood-frame, shingled building surrounded by wooden lean-tos, "was all in a blaze, the flames darting up through the roof a fearful height." Being situated in a crowded neighbourhood of small wood-shingled frame homes and with the lean-tos' roofs rapidly becoming involved in the fire, Broomfield feared the whole neighbourhood would soon be caught in a conflagration. To his surprise and satisfaction, the firemen, working heroically, knocked out the threatening disaster in half an hour.

Looking at the "suburbs" — simply the adjoining neighbourhoods or wards immediately surrounding the old city rather than farflung communities as modern use of the word would suggest — the insurance inspector's eye was not pleased. Broomfield saw little he liked and was glad the Phoenix had little at risk in them. Of Montreal's two major brewery-distillery complexes, Molson's Dow's, he thought the former, along with the same company's iron foundry, a risk his company should get off as soon as it came up for renewal. Dow's he had some doubts about too, but left the final decision on it to the home office.

The Quebec suburb lay to the east, downriver, and was somewhat industrial in makeup because of its harbour front and working-class homes. Broomfield wrote, "I consider this to be the worst district I have yet seen. It is very thickly inhabited. The houses are all low buildings crowded together, all sorts of hazardous trades carried on, and nearly the whole of the buildings are timber and shingles." He showed no envy of the Alliance for their large risk taken in the district. He instructed Messrs. Gillespie & Moffat, the current Phoenix agents, not to renew the £5,239 in policies on Molson & Company's distillery and foundry located in the Quebec suburb. There had been several fires recently in the property, but fortunately they had been extinguished speedily, he noted. In the margin of Mr. Broomfield's original longhand report, opposite these remarks, is a later notation, "Burnt down, eight October,1858."

The other suburbs, although predominantly of wood construction and largely with the wood-shingled roofs that are so effective at spreading fire by flaming brands, were in varying degrees better than the Quebec, or eastern, district. To the north was the St. Lawrence suburb, climbing the lowest slopes of Mount Royal. It was newer, having wider streets and some stone buildings with metal-covered roofs. Broomfield's faith in these wider streets and a few good buildings was not really justified, as later conflagrations in all of these districts would amply demonstrate.

The last of the suburbs was St. Ann's, lying southwesterly in the Pointe St. Charles and Lachine Canal areas. This was another industrial district. It was "more thickly built upon than the other suburbs (Quebec excepted) and fires are often occurring there, but which, owing to the activity of the fire company, seldom extend beyond one or two houses, although the district consists almost wholly of timber and shingled buildings." It was likely the "Queen," an Irish-dominated company, as befitted that ethnic district, that was being complimented on its effective firefighting.

Broomfield could hardly have blotted the ink on the foregoing letters regarding the risks of fire in Montreal when an extensive conflagration occurred that caused him to postpone his October fourth departure from the city and write, on the same day, "It is with unfeigned regret I have to announce to you, that since writing the preceding letters, Montreal has been visited with the most alarming and extensive conflagration that has been witnessed here for many years."

Before piped waterworks, the independent water carter sold water door to door and helped supply the fire engines as well. A premium for the first to arrive promoted both speed and spillage.

"The fire broke out at three o'clock this morning in a building used for roasting coffee in the St Ann's suburb and before the arrival of the engines, which were speedily on the spot, the flames had extended to the contiguous buildings and some delay having occurred in procuring water, together with the circumstance of most of the houses in that district being timber and shingled, the firemen were unable for some time to check the progress of the flames."

The military were called out to help and prevented the fire from crossing Wellington Street by blowing up two brick houses. The fire was eventually extinguished in about four hours, but not until it had destroyed upwards of 100 houses, a large Methodist chapel, a smaller meeting house and other buildings. This included the entire two blocks bounded by Queen, Wellington, Nazareth and Gabriel streets.

Broomfield then apologized profusely for the fact that among the premises lost was Wragg's steam-powered nail factory, insured with the Phoenix for £1,500. It was a wooden building with wood-shingle roof, and the embarrassed inspector had to explain that for some reason its existence had not been pointed out to him — he would have included it on the list to be dropped because of the high risk. Other companies shared the risk. The nail factory cost the Phoenix £1,000, but even with an equal amount from the Alliance, coverage was not enough for replacement of the building. One hundred employees were put out of work. Another company, the County Of Montreal Mutual Office, with many risks lost in the conflagration, was heavily hit for some £18,000, the sort of heavy blow often suffered by such insurance companies not having a wide geographical dispersion of risk. As to the cause of the fire, Broomfield advised that, "The precise cause of the fire has not been ascertained, but it is believed to be purely accidental and not the result of the malevolent feeling of incendiarism."

Disgraceful Behavior In Laprairie

The village of Laprairie, Quebec, was important prior to the building of the Great Victoria Tubular Bridge across the St. Lawrence River at Montreal in 1860. It was the ferry steamer landing where traffic from the metropolis moved to the farm country of the southern counties, the location of the terminus of the first primitive railway to St. John's, and the water route to the Hudson River and New York.

One summer's evening in 1846, at about 8 p.m., a fire broke out in a blacksmith's shop near the railway station. A strong southwest wind off the broad river basin soon communicated the devouring element to adjoining buildings and, in a very short time, the entire village seemed doomed to destruction. The fire continued until five o'clock the next morning, by which time over 300 dwellings had been consumed and perhaps 1,500 people left without shelter. With the exception of about ten small houses next to the river, there was nothing left standing for a depth of two blocks in from the steamboat landing. The various shops, a distillery, and the Scotch and English Chapel were all destroyed.

There was one small fire engine belonging to the village and two to the army barracks. The Protector Company from Montreal, under its Captain Fletcher, responded to assist. They crossed the St. Lawrence to Longueil, then with much delay in getting a team of horses to pull their engines about six miles to the scene, arrived somewhat weary at about 2 a.m.

Coming on the scene the Montreal firemen were dismayed to view a spectacle of general drunkenness, neither soldiers nor civilians being in any condition to work their fire engines. Casks of spirits salvaged from the distillery were standing on the shore. It seems the temptation had been too much to resist. No firefighting was going on. Men were unconscious of where they were or what they were doing. The Montreal newspaper, *The Pilot*, complained they rescued trivial items from the flames when a little intelligent exertion might have saved half a street. "It is melancholy to see the furniture strewn along the beach and women with their children destitute of either food or shelter," said *The Pilot*.

The Montreal firemen, late and blown as they were, set to work and saved the Roman Catholic church and some 40 houses lying beyond it. In spite of this version of the events, military records show the 46th Regiment as receiving praise for their efforts at the great Laprairie fire.

(VII) Three Old Seaports

Saint John's Early Ordeals

Saint John owed much of its strong commercial growth to a strategic location at the mouth of the major navigable river system of the same name. This system gave ready access to a great wealth of timber in the interior hinterland of New Brunswick. Logs and lumber poured forth for export and into the building of the wooden sailing ships for which the port would become world renowned by mid-century. Unfortunately, the St. Johners utilized these same lumber resources to excess in the building of their city in the late eighteenth and early nineteenth centuries. Other eastern Canadian cities had begun utilizing brick and stone to a considerable degree in the more congested central districts, but Saint John did not — suffering painful consequences, as did its namesake, St. John's, in Newfoundland.

No more expert observer could be found than the redoubtable John Broomfield. He arrived by ship from Boston early in October 1845, enroute to Halifax, this being the already well-established transportation link between New England and the Maritimes. He was undoubtedly appalled at what he saw, not yet having been to St. John's, describing the city and its principal suburb, Portland Parish, as being "one mass of timber and wood shingled buildings." Brick buildings occurred, he said, only on the sites of former fires. For this reason, on the north side of the spacious Market Square at the foot of King Street, there was a range of substantial brick buildings with iron doors and shutters, replacing those lost in a recent conflagration. A few brick buildings had just been completed nearby on Prince William Street, on the site of another major fire of several years earlier. In this city of 30,000 inhabitants, there were 12 stone buildings, but even these had wooden roofs and shingles.

The city Broomfield was thus condemning as a bad fire risk posed no great skill in assessment. The record was enough. The fire insurance experience had been bad. In the past 20 years, Saint John had suffered over and over from destructive fires. It might be argued that it was typical of booming cities to build hurriedly and carelessly for today's needs, rather than slowly and solidly for tomorrow's.

The Loyalist city at the mouth of the Saint John River suffered its first major outbreak in either 1823 or 1824, at the end of a boom and inflation cycle in lumbering and shipping. The population had reached 12,907. The fire started on Disbrow's Wharf and swept back along both sides of Prince William Street, a major business thoroughfare near the waterfront, taking out 40 buildings. At this point the lumber export bubble burst and there ensued a gloomy year commercially. Two years later there was a major fire in Indiantown, the part of Portland that served as an important port above the Reversing Falls on the river.

After another decade, on January 13, 1837, Saint John was put into a state of shock by the devastation of most of the business district, 115 buildings being lost. The total property loss was reported to be £250,000, staggering the young commercial city. Many citizens were left totally destitute, losing everything, and without the massive influx of outside relief that became the rule throughout North America a little later. There had been prejudice against insurance, apparently, and little was carried. Occurring in the depth of winter, the fire was likened to the burning of Moscow in later accounts. The experience was traumatic for many citizens and some of them relived it, enduring the even greater conflagration of 1877.

The fire of 1837 started at Peters' Wharf near Market Slip, spreading southward along

A view of the conflagration in Saint John, 14 January, 1837. One of several to occur in the congested business district about Market Square, this fire is consuming the warehouses of South Market Wharf and heading into the city.

-PAC-C41372

the waterfront warehouse rows and for two blocks inland, clearing out important wholesale warehousing and business buildings of Water and Prince William streets to Princess Street and the Carleton ferry landing, about five critical blocks in all. This occurred to the south of the central Market Square and Slip. Only two and a half years later, it would be the turn of the waterfront area to the north of this point. North Wharf Street, North Market Square and Dock Street were burned out at a cost of £200,000. This fire resulted in some attempts by the province to better regulate fire prevention in the city. But the toll continued. Sixty buildings were razed in Portland on August 26, 1841 and that November 14, there was the destruction of the new Saint John Market House in the centre of Market Square. The civic offices were on the upper floor. This fire included a repeat performance through the Water Street, South Wharf area just rebuilt after 1837.

There was something wrong besides the tinderbox construction in Saint John. An explanation comes in comments on the early fire companies made long afterward, these in the form of anonymous veteran fireman's recollections. Casting some doubt on the effectiveness of the fire department of the early 1800s, he wrote that the engines were ill-maintained, dirty and often seized up. When there was a fire, a lot of oiling and hammering was required in order to get the pumps to work. The hose was allowed to become stiff and dry, with cracks and numerous leaks. It was certainly true that leather hose needed considerable maintenance to keep it flexible enough to be usable, and this was a perennial problem everywhere, especially with volunteers.

To sound a fire alarm, the citizen ran to the nearest of five or six bells distributed about town and pulled the rope vigorously. In most cities these were at the fire engine houses, but in Saint John they were on simple towers at convenient locations. Our veteran

Once again, in 1845, fire roars through the Market Square area of Saint John, N.B.

fireman claimed that the Saint John firemen often simply responded to the firehouse nearest the bell being rung, sending out from there a scouting party of the more energetic members to see whether an engine was actually needed. They would relax and have a smoke while waiting for word. Meanwhile, anyone on the scene would use buckets of water from the nearest public or private well. Buckets were reasonably plentiful, as required by the bylaws. Householders were reminded of this requirement in a notice over the signature of Mayor William Black in 1841: "Every householder shall provide two leather buckets, to contain two and one half gallons of water, to be hung in a conspicuous place. Penalty, ten shillings for any month not so provided."

The fire wards had control of the fire scene and ordered bystanders to join the bucket brigade lines. It took a lot of manpower to keep even the smaller, earlier engines supplied if the well was at any distance. In Saint John, as in other garrisoned towns, troops and sailors turned out smartly from their south-end barracks, adding a disciplined body of men to form supply lines.

If a fire occurred in the early evening hours, these being most favourable for a prompt and sizable turnout of firemen and kibitzers, there was some chance of controlling it in an incipient stage, according to our cynical veteran fireman. When a fire reached conflagration proportions, the firemen would abandon the use of engines, resorting to pulling down wooden buildings with hooks and axes. The army sappers and engineers brought powder to blow up buildings in the path of the flames, while the regular troops helped guard salvaged belongings.

There was a brief respite until 1845, when fire struck the upper end of King Street hill, the other end of the business district. This time, in another wintertime conflagration, over 100 buildings around King Square, including the Commercial Hotel, were lost to the flames. The tower of nearby Trinity church caught, but was pulled down in flames, saving the church for another, later fire.

If the earlier fires had not, this latest one drew attention to the antiquity of the fire engines, including the two originals, and the aging, socially-oriented firemen manning them. A public meeting in the Mechanics Institute hall resolved upon the reorganization of the fire department. Young men joined enthusiastically, helping create a new Saint John Volunteer Fire Brigade. They took over the old engines and put them in shape. The city's Common Council officially ratified the new department with its six companies in the city proper and two engine companies in the Carleton section on the west side of the harbour. Each company had an assistant engineer, foreman, assistant foreman and 60 firemen. Five years later, adjoining Portland organized two engine companies of its own, replacing its old Fire Club of 1841.

While the reorganization was going on in Saint John, another fire on Water Street destroyed 40 buildings on July 29, 1845. The new fire companies went to work to improve themselves with a more typical volunteer fire company spirit than the old companies had exhibited. No. 5, the "Silk Hats," were the first to raise the funds to import a modern Smith piano hand-engine from New York. They were also the first to acquire a new hosecart and hose supply. The engine was delivered in October 1850. The other companies followed suit, obtaining the American Smith or Hunneman engines or Montreal-built Perrys.

At first the new companies were handicapped by the limitations of water supply. The first primitive piped water supply system was inaugurated in 1832 from nearby Lily Lake and was being slowly extended through the solid rock upon which the city was built. The fire companies maintained lists of wells, public and private, in their engine houses for suction supply to their engines. This was important in the contest to get the traditional award of one guinea for first water on the fire, as well as for efficiency in the protection of the city. The Saint John Volunteer Fire Brigade now entered upon an era of relative competence and was able to marshal an appreciable firefighting strength in this time of wooden engines and iron men.

For the older towns in the East, the powerful New York built Smith piano engines were a second generation of pumping power. This one served Western Star company of Carleton, N.B.

Prowess In Halifax

As an important British military base, Halifax was a different kind of city from Saint John. This made for a different sort of fire experience.

Halifax was dependent, as it long would be, upon the comings and goings of the British army and navy. In 1800 there were 8,000 civilians to 4,000 military. Around this time, the total tax assessment of the town was $6,000, and local taxes were about three dollars per house. Each householder did six days road maintenance work for the town or hired a substitute at three dollars for the week. There were still stumps of trees in the muddy streets, no sidewalks, and Indians were considered a danger beyond Bedford, ten miles out.

Rum was plentiful from the West Indies trade and there was a gay social life in the 1790s, revolving around the military commander of Nova Scotia, the Duke of Kent, father of the later Queen Victoria. To head off fire calamities there was a chimney-sweeping bylaw to reduce this major fire cause, while smoking was prohibited on the wooden platforms commonly installed in front of the shops. Street paving began only in 1824. Heating was by fireplaces, stoves not yet being in general use.

By 1835 there were five fire engine companies, or as they were designated in Halifax, "sections," of the one overall Union Engine Company. Each was limited to 60 "engine workers." In addition there were the five "fire companies," comprising a large number of the ablebodied citizens and apparently dedicated mainly to protection duties, though they also assisted the engine company and formed bucket brigades.

Water was a problem on the rocky hillside site of the town. Portable hand pumps raised water from the harbour to the wharves to fill buckets for the bucket brigade lines. These pumps were also used thus at various wells. A portable supply was later added by providing a trough and two casks on wheels for interim supply to the engines until the bucket lines were formed. Water was especially scarce in the higher levels, so it was arranged that if a fire was beyond 600 feet of hose run up from the harbour, No. 1 Engine would go to a wharf and relay water through a hoseline to No. 2 farther up the hill.

A fire bell was imported from London in 1822 and installed at the market opposite the town guardhouse. A bigger one of 417 pounds was presented by the Sun Fire Company to Mayor Alexander Keith in 1844 and installed at the central engine house at the Grand Parade. This is presumed to have displaced the army buglers as the primary town fire alarm.

The Axe Company, founded on July 12, 1813, performed rudimentary hook-and-ladder duties, but their strong suit was likely axe work, getting at the seat of fires and cutting away burning parts to prevent fire spread. Each member had to provide himself with an axe or "spudge ," this tool not to be used for anything but firefighting. The company, like the others, took orders from the fire wardens. Its hooks and ladders were kept in the central engine house, and it was only in 1828 that an offer was received and accepted whereby a carter agreed to respond with his horse and truck to carry the hooks and ladders to fires. This was done in return for exemption from statutory road work. Halifax Axe was likely the first ladder company in Canada.

Ladder service improved to the extent that there were two hook-and-ladder carts in 1849, each carrying hook ladders and wall ladders up to forty feet in length, two grapnel hooks and chains, and poles for supporting the larger ladders. A piped waterworks system was in service by this year, the fireplug keys having been distributed about town, and on May 11 the minutes of the Axe Company reported simply and succinctly, "Fire broke out in Barrack Street and burnt two notorious houses of ill-fame."

Halifax had had some severe fires, but none of the repeated conflagrations of St. John's or Saint John. The Phoenix's Broomfield visited the city in October 1845. The population was then 25,000. He observed straight off that it was also largely a town of wooden buildings and shingled roofs, only the numerous government buildings, and occasional warehouses and large homes seeming to be of stone with slate roofs. An advantage was that the streets were regularly laid out and at least 50 to 60 feet wide.

Broomfield was impressed with the fire department, it having a number of good and periodically tested engines, together with a regular corps of firemen in the Union Engine Company. In addition, pretty well every householder seemed to be involved in one of the five protection-type of fire companies, each responding with his bucket, canvas salvage bag and lantern to assist in carrying water or salvaging goods. On top of this there was the large garrison available to assist. The hydrant water system was only in the planning stage at the time of his visit.

Why did Halifax have a better fire record than other cities? There is a distinct impression that a strong influence for order and organization by the military had been the cause of a well-laid-out town, better and enforced fire regulations and a disciplined community. To this may be added the in-depth firefighting resources, civil and military.

Confirmation of Broomfield's view that Halifax had a strong and well-disciplined firefighting force in the 1840s came to him the following year on a return visit, when he witnessed a fierce waterfront fire being confined to the group of buildings in which it originated. Thanks to the military response to town fires, the number of fire engines promptly on the scene was more than any other Canadian city could have mustered in all.

Broomfield said that 14 of them were in operation within ten minutes of the alarm, a remarkable achievement and something not possible to achieve today in terms of numbers of pumps only. A strong force of axemen rapidly cut down a nearby wooden building downwind from the fire. The weakness of combustible soffits under the slate roofs of stone buildings was an Achilles heel and required great alertness to prevent the entry of fire. This was a common problem in the overhanging eaves necessary in snowy climates like Canada's, but was capably handled by the soldiers and firemen in this case.

The training, methods and energies exhibited at this potential conflagration in the dockside warehouse area was beyond all praise, according to Broomfield. The outbreak occurred at 10:30 p.m. in a carpentry shop building in the Long Wharf. A great volume of flame lit up the Dartmouth shore far across the harbour and flaming materials floated on the water as vessels in danger were worked away from the wharves, a very spectacular sight. The local press waxed poetic over this, but noted that the occurrence of incombustible buildings and alert firemen had demonstrated to the citizens once more their value in averting conflagrations.

Nevertheless, the city did not get off scot-free by any means. There was the Granville Street fire of 1831 that wiped out nearly a whole block downtown; the North Barracks Fire of December 24, 1850, that took 100 houses with it; and the previously mentioned New

The Royal Navy in the days of sail carried light portable fire engines aboard. Sometimes they helped out ashore. Halifax had both army and navy engines available.

Year's Day Fire in 1857. This last one destroyed one fire engine and damaged another, while the Halifax Insurance Company suffered such large claims that it required reorganization. Last of the series of blows was one that destroyed the Exchange, Oddfellows Hall, American Consulate, and several other major business buildings in 1861.

Following the fire of 1857, replacement fire engines were ordered from Perry Bros. of Montreal, and on November 8, 1858, a grand turnout of the combined firefighting forces of town and garrison was staged at the naval dockyard, apparently under the sponsorship of the military. The various companies demonstrated their proficiency in contests viewed by the lieutenant governor, the vice admiral of the dockyard, the major general of the army garrison and distinguished gentlemen of the community. Small military-style engines were operated by the Ordnance, Royal Artillery, 62nd and 63rd Regiments, and the naval dockyard. The six town engines included those manned by the sections of the Union Engine Company, named at that time: Rapid, Alliance, Halifax, Resolute, Aetna and Alma. The Axe Company covered the Issuing Stores Building with ladders as their demonstration. The events ended in a sumptuous dinner at the Masonic Hall, provided by the insurance companies and their agents.

The citizens of Halifax were outgrowing dependence upon the military for firefighting. The community fire companies were now established, a vital service to the municipality, and about to become a social and political force to be reckoned with, too.

Quebec Learns the Hard Way

The growth of a strong roster of volunteer fire companies did not take place in the City of Quebec in pace with their development elsewhere in the early nineteenth century. The existence of a paid fire company under the Quebec Insurance Company to protect its own risks tends to confirm the lack of well-developed volunteer protection. Broomfield reported in 1845 that there was unfortunately a lack of an efficient fire engine establishment and good water supply. There were only two or three fire engines and no regular organized fire engine companies, men being engaged casually to work the engines whenever a fire broke out and paid only for time spent fighting it. Broomfield probably included the Quebec Insurance firemen in this category and was rather casual about establishing the precise number of fire engines, as he was inclined to be in other cases, too.

The water supply for firefighting was still very poor in Quebec, except along the waterfront, and there was dependent upon the state of the tide. In Upper Town there was dependence upon wells and several tanks built for the purpose. There were the water carters as well, but their sources of supply were also remote.

Perhaps the Quebecers were lulled by a reliance on the solid stone walls and tin-clad roofs of the central part of the city, but rapid growth outside the walls was producing whole districts of flimsy, poorer-class wooden homes to accommodate the swelling immigrant population.

Like Jenkins Jones 37 years before, Broomfield was not enamoured of what he saw of Quebec as an insurance risk. He saw the two-foot-thick walls of stone in the commercial buildings in Lower Town, with their sheet-metal roof cladding. But he also saw the extremely narrow streets, exposure to the winds off the river, and the wood clapboard weather-covering over the stone. The locals claimed the clapboarding would easily pull off to prevent the spread of fire, but Broomfield was skeptical that it would be done in fact.

The Englishman criticized the overhanging eaves of the Canadian buildings, that allowed snow to slide off but lacked the fire protection of a parapet. He was also advised that, while the tin roofs afforded protection against sparks, the intense radiant heat in a conflagration simply rolled the sheet metal up like a scroll, exposing the boards underneath. Nevertheless, Broomfield felt Lower Town would not be considered such a bad fire risk if it were not for his poor opinion of the firefighting organization.

Upper Town within the walls, he compared to a well-built country town in England, but with the added hazard of the large Canadian woodburning stoves with their long metal stovepipes, needed in winter for heating.

The largest risk on the books of the Phoenix was the Quebec Seminary and attached church, covered for £11,300. He gave some restrictive instructions to the agents on amounts to be taken in insurance coverage in each part of the city, pointing out that the local Quebec Insurance Company had never been able to make a profit on their business and were in difficulties due to the conflagrations of that year.

As should have been anticipated, the suburban powder keg of flimsy wooden houses was touched off eventually. Not one, but two conflagrations roared through these districts in the early summer of 1845. The first fire occurred on May 28, an ideal conflagration day — hot, dusty dry, with strong winds. At 11:45 a.m. the church bells of St. Roch ward pealed the alarm. Under the cliffs of Cape Diamond at St. Vallier Street, fire had sprung up from a large tannery surrounded by wooden houses. Showers of sparks settled on the surrounding wood-shingle roofs and the dry, pine-planked roads. There was little water in the area and few men showed up to fight the fire. The dry grass on the ramparts above burned like straw in the gale.

Troops were turned out to help, and they attempted to make a stand at the government fuel yard at St. Roch Street, which the fire had reached at 3 p.m., but they were driven back to the gate in the city wall. Some fire engines were lost, while others managed to escape. The flames reached very close up under the powder magazine on the bluff, but the troops fended them off, demolishing surrounding buildings and claying up the doors of the magazine, saving the 200 tons of gunpowder from demolishing much of Upper Town.

Eventually there was about a mile of burnt or burning houses extending across the Lower Town St. Roch area. At one point, the soldiers turned their gunpowder to use, using a whole barrel of it to clear a break. Then rain came, unseasonably late, to put a damper on proceedings, but only after the wholesale destruction of St. Roch had driven its thousands of inhabitants to the open fields, with a loss of 20 lives and 1,050 buildings.

It was Upper Town's turn on June 28, at midnight, a house on the ramparts in St. John's Ward bursting into flames on a rising wind, the result of someone dropping still hot ashes on a refuse pile containing straw and shavings. Sparks and blazing shingles flew ahead of the fire on the wind, spreading a conflagration right through the whole ward to the outskirts of town, raging through the night, despite the blowing up of buildings. Again, heavy rain came only in time to wet the smoking ruins. Virtually the whole ward was wiped out, some 1,200 buildings, and 40 lives were lost in this middle-of-the-night fire.

Sources of figures conflict, but at least 16,000 people were burned out of their homes in these two fires, which covered about two-thirds of the area of the city. The central business areas were not involved though. The dispossessed lived in tents provided by the military or in lean-tos in the ruins. Temporary shack homes were put up in backyards by those who could afford them, to provide accommodation for the coming winter. Permanent houses would then be built on the sites of those destroyed. The rows of shacks, equipped with metal stovepipes instead of masonry chimneys, were a serious fire hazard in themselves and a threat to the permanent homes being erected in front of them.

Although any description of temporary building was allowed in order to get roofs over peoples' heads, a new law required all permanent buildings to be of brick or stone, with tinned roof. Two years were given as the limit of use of the temporary buildings. The new laws also required that wooden houses, outbuildings, sheds and fences be whitewashed or painted every five years to reduce their flammability. Lack of paint and the consequent drying out of wooden-cladding on poorer houses has always been conducive to rapid fire spread in the past.

As in the case of most such laws in the early days, there is little indication of how well it was enforced after the shock of the disaster had passed. As to firefighting measures, there is evidence of concrete action this time in Quebec.

Looking across the St. Charles River at the great conflagration of June 1845 in Quebec, with the ruins of the fire of the preceding month in the distance.

- PAC-C79207

Probably the first large loss of life in a fire in Canada occurred with the destruction of Quebec's Theatre Royale. Forty-seven died in the 1846 calamity.

Quebec City must have wakened up a little once again in the aftermath of these two fires, for within the next three years the firefighting service grew to eight fire engine companies, each with an engine and hosereel, organized under a fire inspector, company captains and lieutenants. The captains all had English names, but the makeup of the membership for the companies was four English and four French-speaking. Two were located on Champlain Street in Lower Town and named Faugh a Ballagh and Erin go Bragh, reflecting the ethnic turn of this district. Company members were still paid for their services in each fire fought, captains five shillings, lieutenants three and firemen two. Three of the companies were in Upper Town and five in Lower Town. A hose company, and a sappers and hook-and-ladder company were shortly thereafter in operation in Upper Town. The population was now 40,000.

The City of Quebec had suffered a serious blow in these two fires. Over 2,000 houses were lost. Half a square mile of Lower Town was left a scene of desolation. Quebec had been in decline, no longer the flourishing capital it once was. Much of the shipping traffic had moved on to Montreal with the use of steamships, trade was down and there was a general apathy felt in the city. Yet another tragic blow was struck on June 14, 1846, when 47 lives were lost in the burning of the Theatre Royale, Canada's first large life loss in a single building.

(VIII) Repeated Disaster In St. John's

Built To Burn

The fire history of St. John's, Newfoundland, early in the nineteenth century is an example of the fatalistic attitude toward frequent large fires where there is a weak sense of community. The capital of Newfoundland existed in an atmosphere of impermanence dating from its beginnings as a summer base for European fishermen. Entrepreneurs and workers of the fishing trade persisted in rebuilding for the next conflagration without coming to grips with the possibility that it might be avoided.

When the Phoenix Fire Office wrote their first fire insurance policy in the new World for a risk in St. John's, in 1782, they probably did not have a clear idea of the fire bombshell the rude town represented. It is likely that the first complete description of St. John's as a fire risk was obtained only upon receipt of the report by their inspector and secretary, the actuary Jenkins Jones, whom they sent out in 1809.

The company had appointed as local agents, in 1805, Messrs. Marmaduke, Hart and Company, but limited their business pending preparation of a detailed plan of the town. In response to a request from the governor of Newfoundland, in 1804, for a subscription towards a proposed fire engine and 100 buckets, the company obliged by shipping with their compliments a fourth-size fire engine. (This would have been a very small one.)

Mr. Hart had provided the Phoenix a mixed assessment of the hazards of St. John's in 1807 and that may have prompted the visit by Jones. He said, "The town of St. John's is about a mile in extent running from northwest to southeast. The principal stores (warehouses) are situated in the centre of the lower street. Many of them are contiguous to and some of them are built over coves of water which run up into the town and are intended to preserve fishing boats during the winter. It appears that St. John's is very much within the hazard of one fire, although no accident of fire has happened since the year 1780. There are four or five engines in the town and regular fire companies under good regulations."

St. John's was the first British town in what is now Canada, but it had great trials before becoming a permanent habitation. Long a summer base for fishing fleets from across the Atlantic, permanent settlement was actually forbidden as a policy of government for a long time. Fishermen established their temporary "rooms," or shore-side sheds, for the summer and each year the shanty town was ruled by the Fishing Admiral, whomever was the first ship's captain to arrive in the Spring.

Long years of fighting between the French and English over the fishing ports of Newfoundland followed. St. John's was captured and destroyed in 1696. Peace came eventually, a permanent town growing at the base of the steep hill rising from the fine, naturally sheltered harbour of St. John's. Existing for and because of the harbour and being without much regulation by government, St. John's became a jam of crowded, flimsy wooden buildings, all crammed into the one-mile stretch of shore and little more than two streets deep against the hillside.

It was this picture that met the gaze of Jenkin Jones on his tour of inspection in June 1809. His first observation on writing London was that "it confirmed the utility of leaving nothing unseen." This would have been a good maxim for any insurance underwriter in those times. Here is part of the description of St. John's Jones forwarded to London:

"The body of the town is comprised in the Lower Street, which is from one extreme of the harbour to the other, not much less than one mile and three-quarters long. This street is formed on one side by the stores and buildings on the wharves, on the other by rows of buildings used for retail shops. Although called a street, it is in fact only a crooked and narrow alley, in some places contracted to six feet wide, in others expanding to as much as 12 or 18 feet. This street runs nearly in the direction of the prevailing winds, and is a funnel through which they blow occasionally with great violence.

"The buildings are, uniformly, throughout the town of wood and have the roofs covered in with shingles. There is not a private building of a different description in the whole place.

"St. John's has of late years rapidly increased in population and buildings, and the properties on the water side have become of great value. In proportion with these causes, the buildings upon them have become more numerous, and indeed, a more crowded range of wharfage and stores I have yet seen in no place. Even the narrow coves reserved by government for boats have been encroached upon and every year adds to the evil."

Jones concluded that for most of the length of the town there was no discernible point where a sweeping conflagration might be stopped. There were no substantial brick buildings nor open spaces, and there was the condition typical of wood-shingled towns, in which embers could spread a fire by jumping from roof to roof even if the lateral spread at ground level was stopped. "That whole space is so closely built up, the property accumulated upon it is of such magnitude and bulk, and in many respects of such a quality, of rum, pitch, staves, etc., that the havoc might be dreadful if a fire should ever unfortunately get to a great head," he reported, and concluded, "As a fire risk, it is impossible not to consider the mercantile part of St. John's as one of the most hazardous and formidable kind."

There were at the time three fire companies in the town, with engines for their use. In addition there was the fire engine of the army garrison and, apparently, two or three private engines owned by merchants. Jones said that the engines appeared to be kept in good order and the companies' exertions at fires were very laudable. But he found the town was under very little regulation as to measures of fire safety. Regulations were beginning to be adopted, but the only step which would improve the situation was to require building construction to be of stone or brick. Though this would cost little more, he saw little chance of it. This was likely because the land was leasehold, not owned outright, and no one wanted to build to last.

Looking on the optimistic side, Jones felt that St. John's had the advantage over towns in the West Indies due to the snow cover on the roofs in winter and the probability that the northern seafaring inhabitants would fight a fire more vigorously than citizens in the more sunny clime. However, accounts of great fires soon thereafter suggest that, in fact, firefighting organization was rather ineffectual, or even broke down completely in the face of a severe fire. Jones did not seem to recognize any lack of incentive for pulling the merchants's chestnuts out of the fire on the part of sailors and fishermen living in near serfdom, as was the case in Newfoundland then.

Although Jenkins Jones clearly reported three fire engines and companies in St. John's in 1809, various accounts state that the first organization was the Society for the Prevention of Fire, set up in 1811 at the instigation of the governor, under the leadership of Major James McBraire. The latter was a local merchant who was known to own his own fire engine for the protection of his business premises later on.

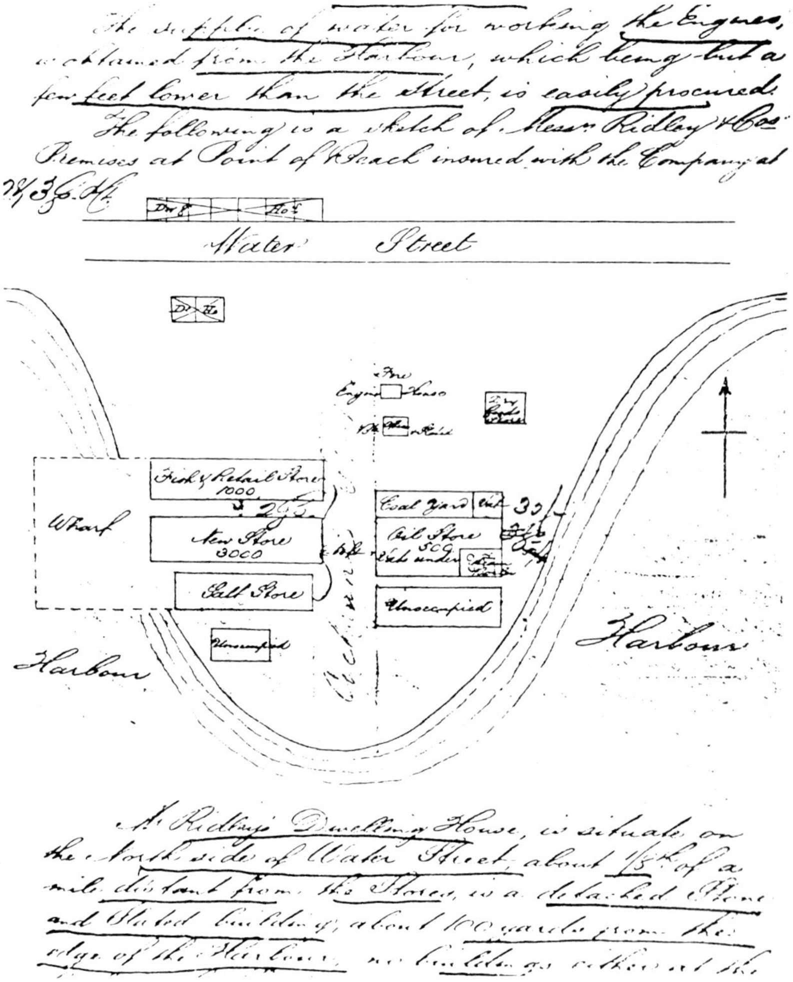

Fire insurance underwriters drew thousands of plans over many years to illustrate fire risks of properties. One of the first was this sketch of Harbour Grace, Newfoundland, by John Broomfield in 1845.

It is possible, as in other cases, that the earlier firefighting organizations were so weak as to have been unworthy of the name, or that they had fallen by the wayside through lack of interest. Another company was formed in 1822 and yet another in 1836, but the total was still only three, including one on the hilltop above the centre of town at Fort Townsend. One of the engines at this time was, of course, that presented by the Phoenix in 1805.

Thus, in 1826, there was an engine company in each of the three wards: east, central and west. This was to remain the basic fire station coverage of St. John's until the 1960s. The garrison engine was centrally located and said to be counted on for important help. Army discipline was no doubt a factor in its value on the fire scene, as was the case in Halifax.

All of these advances since Jenkins Jones' visit were surely the result of the disasters that befell the city, for luck ran out in 1816, and '17 and '19. On the second of February, 1816, a furious fire, fanned by the wind, left 1,000 people homeless out of a population of 10,000. It burned the north side of Upper and Lower Water Street, from Prescott Street to Fort William, having started at the King's Beach.

A Social Alienation

What went down in history as the Great Fire of 1817 was really two fires, two weeks apart. These destroyed much of the business district, including respectively, 130 and 100 houses and destroying food and supplies to cause a shortage through the winter. Attempts to pull down buildings were of no help in arresting the fire's progress.

The winter of 1817-18 was the coldest on record. In the preceding summer the inshore fishery failed, as did the following Spring's seal fishery. From the last in this series of devastating conflagrations, we have an eyewitness account from a participant, albeit a rather cynical one. He was British naval officer William Nugent Glasscock,who wrote his reminiscences in *A Naval Sketchbook*.

According to Glasscock, for a few years the governor, with a small squadron of naval vessels, wintered at St. John's. In each of the first three years of this practice it seemed that there was a major fire. Wrote Glasscock, "On the close of the fishery season, or rather the return of the Paddies (as the predominantly Irish fishermen were nicknamed), an annual fire was as regularly looked for as the coming of the frost."

The town was built entirely of wood. There were vast quantities of fish and seal oils, either in storage or exposed in the open air and being rendered in vats in every direction. Glasscock thought there was no people so liable to be "frost-burnt" as the company of fishmongers of St. John's. When there was a fire, if the governor suggested any precautions or issued orders for the prevention of fire, he would be attacked by the radical press and pamphlets as encroaching on civil rights.

It was one midnight, near the end of August 1819, when the alarm gun of the naval squadron's flagship blasted out a warning, echoing about the steep rocky hills around the harbour. The flagship's night watch had spotted the flicker of flames rising from the darkness of the town. The alarmed inhabitants issued from their homes just in time to hear the peal of the fire bells. Flames spread rapidly through the crowded combustible buildings, the weather having been unusually dry.

The squadron dispatched shore parties armed with buckets, hatchets and hawsers, leaving their small ship's fire engines onboard to defend the vessels. These were moored close to shore, opposite the town and in the lee of the fire. The fire engines played water on the rigging amid the showers of sparks that descended upon them until they were got under way.

Glasscock claimed he was one of the first ashore with the landing parties. He was surprised to see among the tumult, the philosophically passive, whom he judged to be those who were insured, many of them about to be saved from bankruptcy. Then there were the uninsured, these too irresolute or too frantic in their conduct to be of any use. Then again, "the poor, particularly the Paddies, were delighted beyond measure at the prospect of plunder which presented itself and the favourable opportunity now afforded them to retaliate past favours upon their employers." The writer seems to have been well aware of the feudal condition existing in the Newfoundland society of the time. A few years earlier, Jenkins Jones had missed what is only today beginning to be seen as an important factor in the fire picture — social alienation.

The military garrison were no less quick than the navy, forming bucket brigades. Both forces were disciplined and courageous, all the while abused by the unruly crowds. The military's "interference" was questioned. The fire had started in the northeast and the wind was from that direction, blowing it through the town. The soldiers and sailors worked hard to arrest the fiery destruction. Wet blankets and carpets were spread on roofs and walls of houses, but it was all ineffectual.

The military proposed to blow up houses ahead of the advancing flames, but this was rejected, and axes and saws were resorted to. The supporting uprights of buildings were sawn through, but would not pull out. Seamen climbed to the roofs to drop ropes around the tops. The hawsers were then pulled on without result, until the watching fishermen were exhorted to help by their priest, Father Fitzgerald. Down came the houses. The needed breach was made to stop the fire, according to Glasscock, and the conflagration petered out.

In any event, Glasscock said, much property was plundered by the Paddies. On previous occasions, he added, so much loot was secreted away in the local jackass boats to the nearby outports that Governor Hamilton ordered a guard be rowed in boats in the harbour during all fires by the men-of-war crews.

Other accounts give a little more credit to the local forces, saying that the fire spread was stopped at one point by one of the three town engines. The army ordnance had an engine and there were private ones, such as that of Mr. Job, one of the merchants. In this fire, last of the series, over 1,000 people were left homeless and 170 buildings lost.

Too Little and Too Late

After its major fires of 1816, 1817 and 1819, St. John's, Newfoundland, continued through the middle of the century to compete vigorously with Saint John, N.B., for the greatest number of disastrous fires that could be borne without doing much to stop them. Making much the same observation as William Nugent Glasscock had after the 1817 fire, J.B. Jukes, the pioneer geologist in Newfoundland in 1839, commented on the good work of the more solid citizens, the military and the fire companies, but he lamented the element of rabble, those who saw a good chance for plunder when there was a fire. He, too, believed such action to be provoked by what he termed "the system of vassalage" that existed in Newfoundland.

An intense fire which the firemen could not approach to get water on due to the extreme heat, caused great damage on Water Street on June 17, 1833, but it did not rank in St. John's terms as a Great Fire as did the next one, in 1846. The local businessmen attributed the rare occurrence of serious fires during the interim years to the fact that the lower class of inhabitants had all become teetotalers, the ardent spirits having fallen into disuse for some reason. Most fires, they claimed, had started in the grog shops and working-class homes.

In the autumn of 1845, J.J. Broomfield paid an inspection visit to St. John's at the end of his extensive Canadian tour. He gave the city his most scathing report, saying it was "the worst built town I have seen since I left England." However, there were improvements in the offing and he actually wound up on an optimistic note. There were now some seven north-south cross streets along Water Street classed as firebreaks and of 60-foot widths. Broomfield could not judge just how much value they would be in a strong east or west gale. The town was, as in Jenkins Jones' day, a narrow strip along the harbourfront two to four streets deep.

A recent act of the legislature had dictated that the buildings on Water street, the principal thoroughfare, be torn down as each lease expired, all being on leased land, and not allowed to be re-erected closer than 60 feet from the one across the street. Some parts of Water Street were only ten or 12 feet wide, while others already conformed to the new rule. The new buildings also were to be of brick or stone with slate roofs. Some leading merchants were already protecting themselves by replacing their wooden warehouses with iron-shuttered stone ones. Six of them even had their own fire engines, including Baine, Johnson & Company, who had also installed a high-pressure gravity water supply system capable of supplying direct hosestreams. According to Broomfield, the private engines were kept in good condition, but the town engines were old and unfit.

A waterworks company was in process of organization and a gasworks had already been installed. A new fire company was being formed, the firemen to be both paid for their services and exempt from taxes. The Fire Wardens and Company Captains Committee petitioned the board of directors of the Phoenix on January 6th, 1846, for the donation of

One of the early English bedposter style engines, unable to steer, is preserved in Newfoundland, typifying private, insurance and volunteer company engines of the very early 1800s.

two fire engines, saying they would purchase the other two required. They advised that the engine donated by the Phoenix in 1804 was now worn-out.

One of the fire companies, the St. John's Good Intent Volunteer Fire Engine Company, made their own pitch to the board of the Phoenix, suggesting that they were already organized and would appreciate an engine directed specifically to themselves. Their motive is not recorded, but the plea was not considered.

The board of directors responded to the group request by "giving directions that two fire engines of the best and most powerful construction be forthwith provided, to be presented in the name of the Phoenix company to the Town of St. John's for its general protection." The board also hinted to the chairman of the fire wardens and captains of the St. John's fire companies that they should be "periodically exercised and maintained in good order so as to render effectual service whenever the necessity for their use may unfortunately occur." The two engines were duly shipped from the London docks in the *John and Mary* on June 18, 1846.

But they were to be too late. Broomfield, who had carefully examined the buildings of the town, corrected the detailed insurance plans for them and eventually mailed off from London the bill of lading for the two fire engines, was to find himself back in Newfoundland a month later, ruefully viewing the scorched site where St. John's had stood.

More Agony in 1846

It happened on June 9th. A strong gale-force wind was blowing through the length of town from the west. At about 8:30 a.m. a fire broke out, apparently in the west end of the business district. Within an hour the buildings on both sides of Water Street and adjoining duckworth Street were a seething mass of flame. The firemen could not approach close enough to get a stream of water on any of the blazing structures.

One vast sheet of flame, driven forward by the fury of the gale, roared along the two streets the length of the town, moving with such rapidity that the fire engines could not be brought to bear in any effectual way — not that their puny streams would have had much effect. Firefighting operations collapsed as many of the firefighters abandoned their efforts, rushing to their homes to save what little they could of their own belongings. The fire was left to sweep unchecked and uncheckable, clearing away every building in its course to the extreme east end of the harbour, devouring all the works of man, pillars of commerce, hovels, shipping and wharves.

The merchants, complacently relying on the 60-foot-wide firebreak streets opposite the various coves, or slips between wharves, dallied in removing goods from their warehouses. Soon it was too late, as sparks, flaming shingles and larger brands, aided by the withering radiant heat, carried the inferno across the firebreaks as they no doubt would have ones three-times the width. At Warren's Cove and Beck's Cove, in turn, the expected stop did not occur.

According to one view, it was considered that a stop might have been made at Beck's Cove and Water Street but for the injudicious step by the governor of blowing up a wooden building with gunpowder. Apparently an excessive charge scattered flaming fragments in all directions. These flaming bits landed on the oil vats of Bowring Brothers, behind their substantial stone buildings, immediately causing an intense blaze that sealed the fate of the entire block.

Baine Johnson and Company's solid stone buildings on both sides of Water Street were also looked to as a fire barrier, but they too went down before the raging element. And so it went, the length of the crowded town, the gale-force wind continuing unabated the whole day. Broomfield estimated £200,000 loss in goods due entirely to procrastination in efforts to move them. Many struggled to move their goods only to lose them elsewhere as the fire progressed.

This latest catastrophe for St. John's took virtually the whole town. Five persons lost their lives, one in the fire and four in clearing he ruins. Remaining were the outlying sections on the south side, although flaming brands even crossed the harbour to land on the various seal oil rendering plants, but not too many for the employees to control. Also unburned were Riverhead to the west and another outlying section on the top of the hill above the town. The vital commercial heart along Water Street, the better-class homes of New Gower Street and the poorer ones of Duckworth Street were left a forest of chimneys and smoldering rubble.

As in Quebec in 1845, and most other cities of the era suffering similar devastation, temporary wooden shacks and sheds sprang up, often producing a new conflagration hazard of their own. The Newfoundland Legislature hastily regulated new construction, prescribing solid brick or stone-walled buildings with slate roofs on Water and Duckworth streets, and both to be opened up to 60-foot widths. There would also be 11 north-south firebreak streets of 60 to 80 feet.

A waterworks company was also incorporated and an engineer imported from Scotland to operate it. Three years were allowed for the existence of the temporary wooden buildings. The businessmen rose from the ashes with good spirits, built temporary buildings, dickered with the insurance companies over their claims and, with great optimism, looked forward to the resumption of normal trade. The mail steamer from England brought £15,000 in gold for relief and rebuilding, with a promise of an equal amount aboard the next vessel. Major insurance companies facing heavy claims were the Phoenix, Imperial, Lloyds and the Hartford.

Six weeks after the great fire, the two fire engines from the Phoenix arrived. A fire engine house was erected at the foot of Long's Hill, where the Phoenix Fire Company was organized and ready for duty the following January.

Another fire company went by the name Cathedral Fire Brigade and was quartered at Waldegrave and George streets. A fire alarm bell was set up for the town on the corner of Henry Street and may have given the intersecting Bell Street its name. Both of these fire engine companies, along with the garrison engine — no others are mentioned — continued in service until 1876, when replaced by the St. John's Volunteer Fire Brigade. This probably represented the switch from independent companies to a single municipally directed fire department, and lasted 16 years, until the Newfoundland government placed it in the hands of the police.

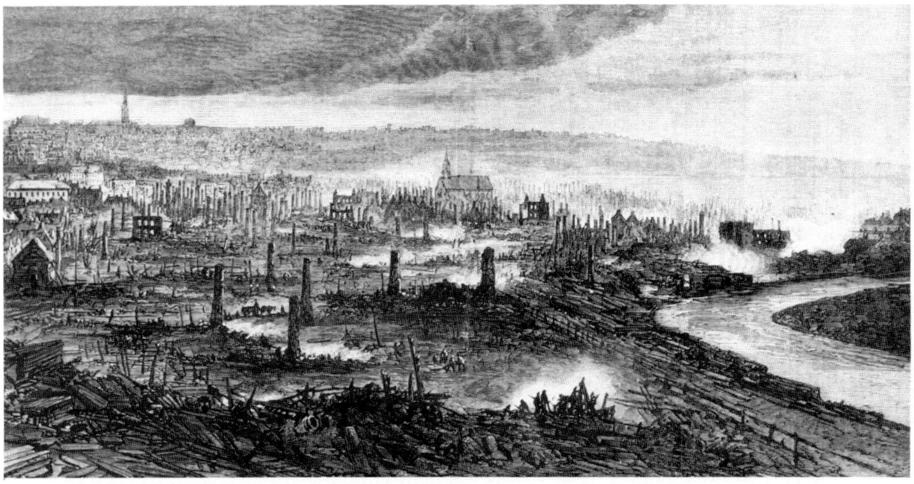

A forest of chimneys is the standard description for the scene when areas of wooden buildings have been destroyed in conflagrations like those in St. John's.

Around the Bay

As the capital and only city, St. John's was sadly under-protected for the day, even compared to a nearby small town. Around Conception Bay was the town of Harbour Grace, old and once a fortified French settlement, its economy naturally based on the fish trade. A population of 3,000 was strung along the shore for three-quarters of a mile in one street. It was a town of wood, but this Water Street was at least 50 feet wide and the houses were mainly on the one side, away from the shore. Three 60-foot firebreaks broke the continuity of the fuel presented by the frame houses. A fire company was in existence and, unlike those in the capital, faithfully practiced with its two engines monthly. One of the two major businesses dealing in salt fish, seal oil and the coal retail trade maintained its own fire engine as well, while the second firm had one on order in England. The major firms were insured with the Phoenix, although there was grumbling about the high rates.

While Harbour Grace seemed the most favourable place in Newfoundland for an insurance company to do business, not so Carbonear, its near neighbour. Larger, with a population of 5,000, this town seemed to have been once prosperous and promising, but now the crowded, poorly-built houses along the narrow streets were in a delapidated state, some closed up and business at a low ebb. It was a picture of a boomtown that had slumped. There were only two brick buildings in the town. The only fire engine received no exercise because there were no firemen. The only water supply was the harbour, while much of the town was on a bluff some distance above it. It was altogether a poor risk, as even the uninitiated would have to admit. Over the next hundred years, however, it was to be Harbour Grace that would suffer the greater fires, suffering two conflagrations to Carbonear's none.

It seems plain that a history of impermanence and a society set in a risky and feudal entrepreneurship supplied the backdrop to St. John's sorry fire record through the 1800s. As always, the point would come when drastic steps would be taken — taken as officials sadly surveyed the smoldering ashes.

(IX) Engines in Ontario

The York Engine Company

Before the first formal Toronto fire engine company was established in 1826, an old pump worked by eight men was carried in a wagon. This may have been the engine of 1818. In 1827 the number of wardens was raised to 15, then in 1838 to 25, each now equipped with a baton of office. Here is seen evidence of the concern over rowdyism and looting at fires in the early days of Canadian towns.

But primitive firefighting conditions continued until 1824, when the new Parliament building burned down. The fire started in the north wing, probably from a defective stovepipe, and spread swiftly. The Upper Canada *Gazette* was sharply critical of the lack of a competent firefighting organization. It complained of "the lack of an able director and an efficient gang to work the engine belonging to this town." It was, the *Gazette* said, a powerful and good engine. It was necessary to point out some self-interest, then firemen could be obtained, the *Gazette* opined. The citizens were to imagine the Market Building and the neighbouring structures all ablaze for this effect.

The loss of Parliament and the editorial castigation of the citizenry may have caused the authorities to wake up a little, like those in Quebec long before. The wheels turned slowly, but on July 1st, 1826, the Upper Canada *Gazette* and the *United Empire Loyalist* said, "We are happy to hear a number of spirited individuals are taking the necessary steps for forming themselves into a fire company under the provisions of an act passed during the last session of Parliament." This was titled "an act to make further and more effectual provision for the prevention of accidents by fire."

This act provided for the establishment of volunteer fire companies in policed towns. The key element was the granting of exemption from serving in the militia, on juries, as constables or in other parish duties. This gave the spirited citizens who volunteered the traditional reward at that time for firemen. The *Gazette* commented that it would be important as well for the police to carry out the regulations when stoves were put up for the Fall season. It noted that the offer of a premium to the first carters to bring water to fires had been found of benefit.

In any event, the first volunteer fire engine company was organized and a new hand-engine purchased and given the name York. The first fire engine house, or firehall, was erected for the firemen on the west side of Church Street, between Court and Adelaide streets. It was of two storeys and brick, with a bell tower. The first members of the company were 38 men from among the merchants and tradesmen organized by Mr. Carfrae, Jr., the instigator of the move and its first captain. He was reelected and remained in this position for six years. The fire engine was described as a 16-man gooseneck engine and was, therefore, probably American-built. It could throw a three-quarter-inch stream of water 140 feet.

The Water Carters

The bucket brigade was soon abandoned as a means of supplying the York fire engine. The local carters brought puncheons of water from the bay on their small horse-drawn carts to fill the tub of the engine at fires. They were paid for this, and in 1825 a system of prizes for the first to arrive was begun.

Water was not that plentiful in York, much of the town's domestic needs being hauled from the harbour, or the bay as it was called then. The use of the independent commercial carters who sold water and carried whatever other loads were offering in town was

The first fire station erected in Toronto was this edifice on Church Street. The adjacent St. James' Church, with the only bell in town, supplied the fire alarm.

common to most communities before the installation of waterworks systems. In York, at first, they received a York shilling for each load. In 1828 the York Fire Company, in conjunction with the Alliance Fire Assurance Company of London, offered better premiums to spur the rapid response of the carters: Three dollars for the first puncheon to arrive, two dollars for the second, one dollar for the third, and 50 cents for the fourth. These were later raised to give even greater incentive.

The carters did not have a very good reputation and were sometimes suspected of arson to develop revenue. Since they were able to respond quickly when they had a loaded puncheon and did not go on foot like the firemen, they often reached the fire before them. This situation gave them the chance on occasion to dicker with the unfortunate victim of the fire to be paid to help salvage his goods. The carters were used in Toronto until 1861. When the first private waterworks system with hydrants went into service in 1843, it was so weak that the firemen continued to require the carters.

York Becomes Toronto

A hook-and-ladder company was organized in York in April 1831. It had 60 members, under a captain and two lieutenants. It went by the name Phoenix, but with no apparent connection with the insurance company of that name. (The Phoenix bird rising from the ashes has always been a popular symbol in fire protection.)

Money was raised by popular subscription to buy another gooseneck engine. It was named Toronto, the Indian name for the place, and No.2 Company was set up to man it. The Church Street firehall was modified to accommodate the three companies. When York was incorporated as the City of Toronto in 1834, the fire department had the two engines, two hosereels, the ladder cart, and 750 feet of hose for each reel. The annual cost of maintaining the three companies was £17,35 shillings.

With the coming of a city government, changes were made. A Fire and Water Committee was appointed, new detailed regulations covered safety for stoves, stovepipes, the carrying of hot coals, and disposal of rubbish, for example. There was provision for one or more inspectors who would check every building monthly for compliance and report to the magistrates. This was a very ambitious proposal. It is not surprising that only one inspector seems to have been appointed, along with three chimney sweeps, in 1839.

A man was retained to wash, clean and oil the fire engines and apparatus, for which he was paid £25 annually. The firemen of the three companies had a common council to handle coordination and joint matters. Members of the council wore a white sash of office and held the same powers of maintaining order at fires as the wardens.

A second hook-and-ladder company was formed in 1835. It took the name Hercules, the city voting £55 for uniforms. Three years later, there was one of those reorganizational shakeups that took place in the volunteer fire departments, usually due to problems of ingrown cliques and rivalries. The city council had been offended by the company captains' spending money on apparatus repairs as well. Council disbanded the companies and reenlisted the men, no doubt declining some of the applications. All appointments and promotions from then on were to be approved by council.

In any case, the smartly uniformed volunteers with their elaborately decorated apparatus were an important part of any civic pomp and pageantry in those days. Marking the coronation of Queen Victoria in 1838, the assembled firemen of Toronto made a fine display parading through town. Afterward they parked their apparatus and engines outside city hall while all repaired to the council room for a "collation."

Battle of the Don River Bridge

Toronto's third fire engine company was founded in 1837 upon presentation of a new fire engine to the city by the British America Life and Fire Insurance Company. A fire station was built for the company at the corner of Bay and Temperance streets.

It was in the Spring of 1837 that the first minutes of the company were written to say, "A fire engine procured by the British America Life and Fire Insurance Company having arrived at 6 o'clock in the evening, the following persons convened at the engine house and volunteered their services to form a company for working her." There followed a list of 22

One of the more famous of the old hand fire engines is the British America, credited with routing Mackenzie's rebels at the Don River Bridge.

citizens. A committee was appointed to develop bylaws and the company was duly constituted, receiving acceptance as a regular participating fire company by the city council on June seventh. David Patterson, who chaired the first meeting, was elected captain. The enthusiastic members responded to fire calls and carried out drills, the terse minutes saying, "Ordered up two carters with water, exercised engine, took up some fines and adjourned."

In the autumn of 1837, rebellion was in the air and the community was tense. At one a.m. on December 5th, the fire bell rang, turning out the volunteers. It turned out to be anticipatory; the followers of William Lyon Mackenzie were said to be approaching the city to burn and pillage. The firemen went home. One house on the outskirts had been burned.

On the seventh a meeting of the company was held. It was resolved to form a volunteer militia company from the fire company. The members were determined to protect their homes. The company thereupon called on the mayor, who was in command of the defense of the town, to offer their services. The mayor asked that they stand by with arms, engine, cistern and equipment at the Market Square.

The fire cum infantry company exercised on the square until word came that seven or eight hundred rebels were entering the city by way of the Don River toll bridge and had commenced firing buildings. The mayor dispatched the British Americas, the company setting off with great spirit, muskets in one hand, the engine-hauling rope in the other.

On arrival at the bridge, Captain Patterson ordered the engine into action, but on reconnoitering soon found that the buildings were beyond salvation except for the tollhouse, and that the bridge had been set on fire. The firemen tore up the planks of the bridge floor to douse the flames with buckets of water, then headed home, arriving back at the market at 9 p.m. As for the rebel force, these had disappeared before the British Americas came in sight. The heavy rumble of the wheels of the advancing engine and its trailer cistern had frightened them into believing that the artillery was arriving.

At midnight some of the company were turned out to the same area in company with the mayor, but it was a false alarm. The fire militia remained on duty at their engine house for some days after that before standing down.

Little Water - Big Fires

The first overall fire chief of a multi-company fire department in Toronto seems to have been Thomas Harris, who served from 1837 to 1841, likely on an elective basis. His assistant was Robert Beard, his successor and holder of the top position until 1851. These and other officers' jobs were apparently annual appointments based upon election by the members and confirmation by city council, as has been the case in many municipalities down to the present day where the service is volunteer.

Officers or whole companies could take advantage of the annual resignation and election time to register complaints and to pressure for changes. Similarly, the city could take this opportunity to get rid of troublemakers or defuse rivalries by juggling company memberships. In 1849 Chief Beard, his two assistants and some of the companies resigned and accepted reappointment upon a reorganization of the department.

Beard took command of an expanded fire department, having a hose company added under the ambitious name of Niagara. A fourth engine company had been added in 1842, called Victoria. More money was evidently being spent, for a specific tax of three pence on the pound was set to maintain the department.

Perhaps the problems leading to the reorganization of the Toronto Fire Department showed themselves in the conflagration of 1845, the city's first great fire. Water supply was a sore point with the firemen. They threatened to resign over it in 1848, when they claimed there was only enough available from the hydrants to supply one engine. There were even threats to march on the water supply plant. The British America Company had criticized the privately-owned system with its sparse hydrant coverage as a "so-called waterworks." When the historic Rennie's Tavern burned early in 1848, there was no water available from the

Church Street hydrant, the engines having to draft from the bay. It was said that the owner of the system turned out to fires to complain to anyone who would listen that the city fathers were getting all the water they were willing to pay for.

The conflagration of May 9th, 1845, started in an outbuilding of tinsmith William Musson, located between Church and Yonge streets on King at 2:15 p.m. It spread rapidly to a warehouse containing oils, groceries and spirits. The last included 100 barrels of whiskey, which proceeded to explode around the neighbouring vacant lots. Water supply was inadequate due to the garrison not turning out to help, as was customary. They were confined to barracks. The reason is not recorded, but there might have been some connection with the 100 barrels of whiskey rocketing about the area.

Toronto City Hall was burned in another major fire in 1849. St. James' Church and other buildings in the area of King and Nelson streets were also consumed. Witnesses described the helplessness of the firemen as a blazing shingle soaring on the updraft of the fire lodged in the louvers of the church belfry, out of reach. The steeple of this fine edifice took fire, and the flames extended downward to engulf the whole church. This not only destroyed one of the city's finest houses of worship, but also the fire department's main alarm bell. Ringing the bell of St. James' had long been the call to action for the volunteers. For a long time it was the only bell in town. Every citizen knew that keys to the church were at the police station and other city buildings when needed. The firemen also had a setback at this fire when one of their fire engines broke down.

These two conflagrations introduced Toronto to an era of bad fires that extended through the 1850s. Buildings were getting bigger and more crowded in the centre of the city, while the technology of the firefighting service was lagging behind. The puny streams of water were not capable of checking the intensity of burning in the larger amounts of combustibles. The big fire of 1849 burned out 15 acres and was really stopped in the end by a change of wind and a downpour of rain.

Smaller Towns

Through the middle of the nineteenth century the provincial towns of Ontario began to develop with the agricultural prosperity of the southwest, the increasing population and the industry that followed. There had been Loyalist immigration in the previous century, but not in the concentrations that built large towns overnight in the Maritimes. The spurt in Upper Canada came with the European influx of the 1800s.

Cobourg, not far east of Toronto, started up a fire engine company in 1832, fairly soon after the capital, and it appeared at the celebrations staged by the Toronto companies in 1838 for the coronation of Queen Victoria. The Cobourg engine was named Protector and inhabited an old shed at one side of the market. A second engine company, with the name Red Rover, and a hook-and-ladder company were soon added. These companies were composed of 40 to 60 of the most active and adventurous young men about town. Eventually Cataract Engine Company joined them. There was also an unlimited bucket brigade, not so exclusive, taking in pretty much the whole citizenry. It is a tempting conclusion that new fire companies may often have been inspired as much by the desire of more men to belong to a fire company than the existing ones could or would contain, as by a perceived need for more protection in the town. One of the early companies in Brantford was said to have been organized by the Y.M.C.A., possibly an astute move to better compete against the fire companies' glamour for the loyalties of the young men about town.

In 1845 Hamilton was a small town containing about 5,000 inhabitants, situated on high ground about a mile from the shore of Lake Ontario. It was laid out with very spacious streets, one being 112 feet wide in its narrowest part. Broomfield found Hamilton to be rapidly rising in importance, with a great number of new buildings going up and about the same proportion of brick, timber and shingles as in Toronto. All the new houses were being built of brick with tin-covered roofs.

There was no piped water supply for firefighting, only what could be obtained from wells as yet. "There are two fire engines," Broomfield said, "but no regularly organized fire-police. I am told that fires rarely occur here." The leading merchants were Buchanan, Harris and Company, on King Street. For their large property, they had two small portable fire pumps and a central well with a pump.

The earliest records of a Hamilton Fire Department are dated 1854, when there were four engine companies, one hose company, and one hook-and-ladder company. It is unlikely such a number of companies was born overnight. There are indications that the department went through disbanding of companies with subsequent reorganizations as the result of altercations with city council and attempts to rejuvenate poorly functioning companies. It may be assumed that 1854 marked a reorganization of previously existing units.

The first fire engines in Hamilton were said to have been left lying in the open without stations to house them. Householders were required to maintain a barrel of water for firefighting purposes. A system of street tanks was established later for supply to the engines. The department of the fifties had over 500 members. In addition to these, the Great Western Railway shops had an employee brigade of 300 men in four companies.

Ottawa was a late starter among major cities — in 1848 a village of lumbermen called Bytown, after Colonel By of the military engineers who built the Rideau Canal system starting at this point on the Ottawa River. It was then that our first records of a local firefighting organization appear. The origins of the fire companies are lost, but in 1848 there were two volunteer engine companies, named Mutual and Alliance, as well as a hook-and-ladder company. It has been suggested that perhaps wealthier citizens with substantial properties to protect had supplied the engines. It is likely that the leading businessmen had a hand in seeing the fire companies organized, however, the names of the two companies suggest the probability — though by no means confirming it — that the fire insurance companies of these names donated some or all of the funds.

Ottawa favored the Montreal built Perry engines. This one later saw service in Perth. Such hand-me-downs were common.

The executive of the Halifax Union Engine Company in the late 19th century. Their offices were positions of considerable standing in the community.

3

TRANSITION (1840-1880)
(X) Signs of Unrest

By mid-century the volunteer fire companies had grown considerably in number and capability, but they were now showing an objectionable side, too. With numbers, social position and a hold on a vital community service, they threw their weight around politically, or lost sight of their duties in the heat of keen rivalries. They strained the patience of citizens and officials alike in the major cities through the fifties, until their power was swept away by a piece of technology. This was the steam fire engine or, in some cases, an equivalent in the form of a more powerful hydrant water supply system.

The steamer appeared on the scene in the late 1850s, its technology spurred by the demands of frustrated city officials in large American centres where the misdemeanours of the volunteer firemen were extreme indeed. The steamer simply was able to pump as much water as one hundred or more men. It was the means of replacing a large part of this power group.

Relieved of the need for large numbers of men to pump water by muscle power, the city fathers were free to cut their ranks to a fraction and even pay the new force for their services. In a gradual transition, varying by city, the volunteer services changed to an on-call force,paid a modest sum for response when there was a fire. The response was on the same basis as the volunteers, but under pay and control by the municipal administration. With the advent of steam fire engines, horses became necessary. Maintenance of the engines and horses required supplementing the on-call men with a few full-time employees — the first professional or career firefighters.

Problems with the volunteer companies and reasons for their replacement differed in the various cities. Sometimes the volunteers welcomed the powerful engines for their great fire-killing ability. Eventually the on-call forces would pass too, the entirely professional firefighters succeeding them. Smaller communities, of course, retained volunteers, and some 3,000 of them rely on such community-service-minded citizens still. However, for the cities of the 1850s, the freewheeling volunteers had become too rambunctious and disenchantment had set in.

A Power Struggle in Halifax

The Halifax volunteer fire companies were a strong force in 1858, at the time of the garrison engine trials. In April 1860 the roster included the six divisions of the Union Engine Company, each equivalent to a separate company elsewhere, and the Axe Ladder Company. They were augmented by the several engines and ample manpower of the military, as well. The fire record of the city had been good for some time. But luck changed with the serious fire of New Year's Day 1857, when many of the firemen were out of town, the water supply proved weak and two fire engines were burned. The press muttered about the need for a paid fire department.

During 1860 a confrontation arose between the city fathers and the fire companies. It was an example of the usual and inevitable power struggle between the fraternal-style fire companies and the politicians. The Union Engine Company and the Axe Company resigned on May 8th, 1860, in protest against a council resolution asserting right of approval over the supply of apparatus to the Engine Company. The city was paying about £1,500 a year towards the company's maintenance. The 100-odd firemen turned in their equipment on 48-hours notice. This was an example of the tradition of high independence by volunteer fire companies born in the do-it-yourself colonial days of North America. It is still in evidence among Nova Scotia's proud, capable and independent volunteers.

Undeterred, the council promptly appointed a new volunteer engine company of 137 men, effective May 21, and set up in seven divisions under a captain and 14 lieutenants. The now disenfranchised old firemen also reacted immediately, reforming themselves as the Union Life and Fire Protection Company, to function as a salvage and auxiliary force, somewhat like the earlier fire protection companies, these evidently having gradually fallen by the wayside. Later dropping the words "Life and Fire," the Union Protection Company continued in existence in apparent harmony with the replacement U.E.C., and on beyond the latter's demise on the coming of the paid department in the 1890s, until 1919 in fact.

In the fifties both Saint John and Halifax tried the organization of boys' companies to put the eager youth to work helping on the periphery of serious firefighting, but these appear to have been short-lived.

The hold over the community by the influential and socially important volunteer firemen of Halifax was cut when the mayor and an alderman journeyed to Boston in December 1860. There they investigated and purchased for immediate delivery two of the recently perfected and revolutionary steam-powered fire engines. Now the work of many firemen could be performed by a machine that produced much more powerful streams of water and never tired.

Fisticuffs in Ottawa

By the early 1850s the Ottawa hand fire engine companies had grown to four and the old engines had been replaced with the fore-and-aft Perry engines, the Montreal type just made famous by winning a gold medal at Prince Albert's Great Exhibition. Rideau and Chaudiere had their own engine houses, while Ottawa and Alliance (later named Queen) shared quarters at By Ward Market. This provided a station in each of the three wards, but the Ottawa and Alliance did not get along, so a fourth station had to be built.

In 1854 the three-man fire committee went to Montreal to study its fire department, returning to recommend Ottawa follow the Montreal system. It is unclear what it was they wished to emulate. It could have been a hydrant system,with sufficient pressure to make the hand fire engines unnecessary, or the appointment of an overall chief engineer to coordinate and control the companies. In any case, the first chief, John Langford, was appointed the following year. The waterworks system came much later, after much bickering.

The rivalries among the Ottawa fire companies were reputed to be intense, fisticuffs and black eyes being the order of the day at the annual 24th of May pumping contests. Later on, in 1866, the Chaudiere's 50-man team travelled to an international pumping contest in Prescott, accompanied by their powerful but deceptively small Perry engine. This machine was derided by the principal opposition, firemen from across the St. Lawrence River at Malone, New York.

The test was apparently to be the common one: the maximum distance a stream of water could be thrown. Each team would usually get two or three tries. No duration was involved. Wagers were placed and the Americans were taken in. According to the press of the day, the Chaudiere engine was in first-class order mechanically and the pumpers in good form. "They beat the big brass fitted Yankee by a full 30 feet and the Ottawa and Prescott people collected handsomely on their bets."

In Full Flower

The reorganized Saint John Volunteer Fire Brigade showed the beneficial effect of the transfusion of younger blood and a new enthusiasm after the series of great fires in the preceding decade, when, in 1853, they first formally appeared on parade. The newly outfitted companies, in all their parade splendour, drawing their new piano-type engines, turned out for the sod-turning ceremony of the European and North American Railway project on September 14, along with visiting out-of-town companies. Such nattily attired and equipped fire companies, using theme decorations, were, along with military units, the main attractions on most ceremonial occasions of the day.

In this parade were included No. 1 Engine Company of Portland, with a hosecart and band, and the private 120-man company and engine of Robert Rankine and Company's

The grand arch was a standard part of any great public celebration, as was a ceremonial dress parade of the fire companies. Here is Engine 2, Saint John, with hand engine and hosereel, ready to greet the Prince of Wales.

sawmill and shipyard. Next came the Saint John engine companies : Wellington, Union, Cora Linn, Emerald, Fives' Top Hats, Faugh a Ballagh, with Western Star and Prince of Wales from west of the harbour, the Hook and Ladder and the Fire Police Company, the last founded in 1850. Each company had up to 70 men on parade. Following them were delegations of 25 and 40 firemen representing Fredericton's Engines 1 and 2, who brought a splendid hosecart. In the parade, horses drew the engines, as often was the habit on such occasions, but in Saint John possibly it was more in deference to the many steep hills.

The Top Hats of No. 5 seem to have been the leaders of the pack, outshining the competition with a fine brass band imported for the occasion from Boston, their New York engine beautifully decorated and escorted by "black boys and men in Eastern dress, white robes and turbans." The Hook and Ladder Company wore blue jackets, black pants, and black-glazed top hats with H & L in gold on the bands. The others had similarly elaborate dress uniforms, all provided at their own expense.

When the Prince of Wales made his triumphant tour of Canada in place of his mother, Queen Victoria, in 1860, the Saint John companies paraded again, helping to glamourize the occasion, just as did firemen in various other cities and towns. When in Carleton, or West Saint John, the prince was taken about in a carriage drawn by the firemen using their engine-hauling ropes. No. 5 Company erected a great welcoming arch of evergreen boughs across Germain Street at their engine house in the fashion popular in Canada for such occasions.

The Saint John companies, in these peak years at mid-century, had their engine houses beautifully fitted, their rooms decorated with good English carpets and expensive steel-engraved pictures. Here they gathered in the evenings to play cards and games, or read the latest magazines, but bad language, wagers and liquor were strictly forbidden. No. 5 even had a gymnasium, something modern fire stations are coming to again.

Each company staged a ball during the winter season, with dancing and a supper. No tickets were sold, the company footing the bill and limiting the affair to the members, their honoured guests, politicians and those others who supported them. As elsewhere, the Saint John companies were an important social element in their city, in this case for some fifteen years between the overhaul of the floundering old department and the change to a smaller paid department when the hand-powered engine became obsolete.

Young men, not old enough to be firemen, attached themselves unbidden to the companies and were a bother, not being under any sort of control. They had a hangout at a place called The Friary on Charlotte Street, and they called themselves the snowbirds. Most of them became firemen in due course and turned respectable. Fines and threats of expulsion kept the members in line. On the whole the volunteer firemen of 1850, technology aside, were not so different from their descendants in many parts of Canada today.

Thomas Smith, a ship chandler, was the fire chief in Saint John and stood for little nonsense. There was the city council's guinea for first water on the fire to spur competition, but honour was as much at stake in this important race. When the first fire hydrants on water mains replaced drafting suction sources and the hosecarts could be run without their pumping engines, they were fitted with but one hose outlet. As each hydrant could supply water enough for two hoselines, a wye connection was attached before the first hoseline. It became a common trick for the first company on the scene to omit the wye, thus blocking off the rival companies from the closest hydrant to the fire until they had achieved the first show of water. If two hosecarts arrived at once, there would be a scrimmage over the closest hydrant, a few strong words and shoves, at least. Chief Smith took a dim view of this and cracked down. He once knocked down one man with a blow of his speaking trumpet for swearing at him.

During the time of this keen, competitive group of fire companies, Saint John was free of the repeated conflagrations that preceded it. How much the more dynamic fire companies were responsible for improving the record of such a hazardously-built city is hard to say. The waterworks system with hydrants, brought in on a limited scale in 1832 as a

Both monetary rewards and pride sometimes led to fisticuffs among the volunteers of old. The chief's trumpet served to restore order in more ways than one.

Late 19th century portrait of Engine and Hose No. 3 in Saint John, with fulltime drivers and paid-on-call firemen under their own foreman.

private company, had water available to consumers two hours each morning, but was greatly expanded between 1852 and 1855 after being turned over to the city.

The muscle power of the large companies of pumpermen was no longer required as water under hydrant pressure became available, and was totally eclipsed when the first steam fire engines arrived. The heyday of the real volunteer firemen ended in Saint John in 1864, when an on-call paid force replaced them. However, the volunteer salvage corps fire-police would operate for well over another century in their unique property-protecting role.

Keeping the Lid On

At the beginning of the 1850s, Toronto, with a population of 30,000, had a smaller fire department than the older, more crowded and combustible cities of the Maritimes. In 1851 there were only the four engine companies, York, Toronto, British American and Victoria; the two ladder companies, Toronto and Hercules; and the Niagara Hose Company. These were all under the command of Chief Engineer James Ashfield, elected that year even though he had declined the nomination. Ashfield had risen through the ranks from fireman in 1839 to company captain and assistant chief.

The department grew, older engines were replaced and company memberships were shuffled about by the chief to stop developing frictions during the decade. In 1857 and 1858 there was a rash of serious fires; incendiaries were believed to be afoot. The situation was considered serious enough that rewards were offered and a temporary additional fire company maintained in service for about a year under the name Independent. During the fifties the major fire station was on Court Street, housing No. 1 Hook and Ladder, and Nos. 1 and 2 Engine companies. The Mechanics' Institute was on the second floor. This building later became the police station.

No. 2 Engine Company, the "Torontos," were on the outs with the chief in the latter part of the decade, but had the support of Frank Howard, manager of the Royal Insurance Company. With their prize money and profits from excursions and picnics, they bought themselves a first-class folding brake piano engine from Cowing and Son of Seneca Falls, New York. It was a deluxe job that could throw three streams, and it carried a top-mounted reel of hose in the later American style. This company then occupied quarters on Yonge Street opposite Elm, where they had a well-fitted reading room and a bunkroom. This was a great social centre, where the members met to smoke and play backgammon, drafts and whist. The bunkroom suggests that the company kept men on-duty at night for rapid turnouts, a practice some volunteers serve on rotation still.

No. 2 Hook and Ladder was converted into an engine company by Ashfield when the two ladder companies began getting into fights in 1853. It became No. 6 Engine and was named The Provincial for the local fire insurance company that presented to the city, in 1852, the famous Perry engine that had won the gold medal at the first World's Fair in London, England, in 1851.

Chief Ashfield had trouble with the fire companies' penchant for running full-tilt along the wooden sidewalks to and from fire calls, spurning the muddy, rutted streets. He split the hose company in two in 1860 for doing this and to better manage its large numbers. One half of the company became No. 2 Engine Company, the former members having finally resigned en masse that year, presumably over their differences with the chief. The practice of running the engines on the sidewalks had been forbidden by the city under penalty of a fine two years earlier, after Meehan of No. 3 was run over and killed by his engine in such an episode.

At its peak, about 1858, the all-volunteer fire department had some 263 men in ten companies, one of which, the Phoenix, had an auxiliary "boys company." Ashfield, the department chief, began receiving an annual stipend of £250 in 1857. At the Provincial Exhibition of 1858, there was a muster of the ten city fire companies supplemented by

Three fire companies were housed in the Court Street Station in Toronto, Phoenix No. 1, Rescue No. 2 and Hook and Ladder.

visiting delegations from fire companies of Hamilton, London and Cobourg. All paraded to the accompaniment of Mullaney's Brass Band. In 1861 more help from the city came in a new fire department bylaw that provided for the financing of one hook-and-ladder, nine engine, one protection and two hose companies, with a grant of $800 to each for equipment and maintenance.

The municipal authorities had considered the attributes of the recently perfected steam fire engine as early as 1858, but it was not until 1861 that two Silsby rotary engines were acquired at a price of $6,000. The insurance companies contributed one-third of the cost.

The next year was the fateful one for the era of the volunteers in Toronto. There was the inevitable phasing out of the large numbers of volunteers in favour of a much smaller on-call paid department, and the retirement of the hand tubs in which they had so much pride. Nevertheless, the old volunteers still turned out by habit to the stations on the sounding of the alarms for some time. They had to know what was going on.

Fire alarms were still by the primitive means of running to the nearest fire engine house or alarm bell tower and ringing the bell manually. There was no watch maintained in the engine houses, with the result that alarms were sometimes seriously delayed. The

A formal company photograph shows plainly the pride of membership in Fredericton's Hook and Ladder Company in the old days.

meagre private waterworks system had its 85 hydrants, of which a dozen were considered useless because the mains were too small. In 1862 some 28 street cisterns were built to supplement the hydrant supply.

The on-call paid fire department consisted of one chief engineer, Ashfield, and 40 men who worked at other jobs but were paid to respond to fight fires. The department was reduced to two stations, those at Bay and Temperance and on Court Street, housing two steamers, two hosereels, and the hook-and-ladder truck, with three hand-engines in reserve.

The volunteer hand-engine companies carried on in their customary ways in many larger Canadian cities until the sixties, but the handwriting was on the wall for most. Political pressures, loyalty and sentiment could not stop the inevitable decisions by city councils to adopt the modern technologies of piped water supplies and powerful steam pumps. Montreal, Saint John and Toronto were already out of the large-scale volunteer era and relying upon smaller forces of paid firefighters by the middle of the decade.

The primacy of the large, all-volunteer fire companies in the cities was fading, though the volunteers would never disappear in the small towns. Well over a hundred years later, some 80 percent of Canadian firefighters would still be fully volunteer or paid only the most modest stipend for their vital service. In some areas of the country, competition for top performance in contests, quality of equipment and appearance on parade would still be keen, but between departments more than between companies.

In the Eastern United States, the retired volunteers in the big cities formed veteran firemen's associations, continuing on with their annual muster contests using their old engines. Such activity was relatively rare in Canada, however.

(XI) Boisterous Times in Montreal

Company Rivalries

As Canada's commercial and transportation hub, the key city of Montreal steadily expanded under the pressure of new population streaming upriver from the immigration station below Quebec in the early and mid-nineteenth century. By the forties there were in operation, to protect it from fire, nine engine companies, a ladder company and, for a time, a property protecting company. There were also several private fire engines retained by business concerns.

With its more cosmopolitan population, rivalries among Montreal's fire companies existed on ethnic lines, as well as over the usual pride and monetary rewards in being quickest at the fire. The keenest rivalry for response to fires was between the Union (located in Victoria Square near St. James Street) and the Protector (stationed at Place d'Armes); the two were located at opposite ends of the principal business district. Trickery was often resorted to for supplementing skill, but most of the time it worked to the advantage of the citizens being protected.

The Union, probably the most fractious of the lot, due in no small part to the presence of Fred Perry, also had a rivalry with the engine company named Montreal over pumping prowess. This resulted in the building of an engine in 1850 by the Perry brothers, George and Alfred, cabinetmakers by trade, and a great contest held in Place d'Armes. Captain LePage of the Montreal also built some engines, but it is not known whether this contest involved one of these or one of the Hunneman engines. It was to be a typical splash test, but a vertical one, the winners to be the crew that could reach higher on one of the towers of Notre Dame Church, the tallest structure in town by far.

The effort was inconclusive and the result hotly debated. It was thereupon decided to settle the issue by a washing contest. This involved the tub or body of each engine being partly filled with water and both engines pumping through a hose into the other's tub. The object of the contest was to pump faster than the opposing crew, so as to fill and overflow the rival company's tub, thus washing it. In this case, the battle was close and broke up in confusion, with the Montreals accusing the Union of pulling the Montreal hose out of their tub to keep it from filling up.

Both the Perrys and LePage went into the fire engine business after this, but with the Perrys becoming Canada's only significant manufacturer. Under the name George Perry & Brothers, they had sales from coast to coast, working out of a shop on Craig Street.

Some of the original engines were of small pumping capacity and were getting old. In the late forties there was a consolidation of companies, and two new Hunneman engines from Boston were added. The roster of 1851 shows seven companies; the combination hook-and-ladder-and-hose company on Craig Street, and the engine companies under the names Montreal, Neptune, Voltigeurs, Queen, Union and Hero.

Approaching mid-century there was evidence of increasing preoccupation among the fire companies of Montreal with their rivalries and flexing of political muscle, although this paled by comparison with conditions in the major American cities. There were differences in Montreal between the French- and English-speaking companies, but most serious altercations were between the Orange- and the Green-dominated companies, the Protestant versus Catholic feud brought over from Ireland.

There are many tales of American fire companies interfering with the municipal political process, as well as fighting among themselves while buildings burned.These conditions were to become a major factor in the move of some American cities to replace the volunteers with full-time career firefighters. It was the steam fire engine that made this possible by eliminating the need for large numbers of firemen for hand pumping. In Canada conditions generally were not so extreme, but city councils became fed up with having to

Alfred Perry, a big, adventurous and much involved young man about town, was a logical leader among the more fractious volunteers of old Montreal. Here he is in his formal turnout and his Union Company's engine.

handle their fire companies with kid gloves. There was, however, one outstanding case of violent interference with the due process of law and government: the burning of the Canadian Parliament Building in 1849.

A curtain-raiser to the disgrace of 1849 was a legal action by firemen against the mayor and fire committee of Montreal in the preceding year. Military preparedness was of some concern because of the Oregon Boundary Dispute, so the mayor, James Ferrier, decided to organize the fire department into a reserve army battalion. He became the colonel, while members of the fire committee and the company captains were to be commissioned officers. He neglected to consult the firemen on this.

Indignant at being told to shoulder the musket — in their minds simply to serve the interests of a privileged elite — Fred Perry and some other firemen refused to sign up or take the oath. Perry certainly had bitter experience serving in the militia as a volunteer during the 1837-38 rebellion. For their refusal, the objectors were thrown out of the fire department by the fire committee. Turning out to fires and finding his honoured position as branchman of the Union refused him, Perry obtained a court order restraining the mayor and fire committee and ordering reinstatement of the men. The judgment was based on the exemption of firemen from military service. This action soon caused the demise of the fire battalion, according to Perry. The victory is unlikely to have enhanced relations between the firemen and the established authorities.

The Burning of Parliament

When describing the burning of the Canadian Parliament Building during the Rebellion Losses Bill Riots of 1849, most historical accounts simply state that it was the work of a disgruntled Tory mob and let it go at that. The Tories were put out over the signing of a bill to pay for damages done in the Rebellion of 1837-38 because it appeared some of the money would go to individuals who had been rebels. Montreal was the capital of the United Province of Upper and Lower Canada at the time.

The real story is that the mob that marched on Parliament to drive out the members, steal the mace and fire the building was led by none other than the well-known firefighter, hero and now captain of the Hook and Ladder Company, Fred Perry, supported by the members of the Tory, Orange- and Irish-oriented fire companies.

Like many of the other English-speaking firemen, Perry had fought in the rebellion on the side of Queen and country. After his service in the militia he had been shortchanged on his pay, and the government had reneged on the promises of land grants upon demobilization. Thus, when it appeared that the traitorous rebels would now receive money for losses from fighting they had started themselves,men like Perry were aroused to a violent state. They were egged on by strongly partisan newspapers like the Montreal *Gazette*.

The parliament building of the legislative assembly of Canada goes up in flames in April, 1849. The firemen declined to put a drop of water on it.

At a protest rally on the fateful night of April 25th, under the leadership of Hon. George Moffat and Augustus Heward, a petition for the recall of the Governor-General, Lord Elgin, was decided upon. The firemen, among others, were not satisfied with this and called out reinforcements by the simple expedient of sounding the fire alarm. The church bells echoed the call as prescribed, and soon the crowd thickened with the arrival of men in the red Garibaldi jackets of the fire companies, with their heavy belts, long iron hose keys and torches.

The leader among the Tory faction in the fire companies, Captain Perry of the Hook and Ladder Company, was called forward to speak. Always a man of action, Perry proclaimed that the time for petitions had passed and that if the men present were in earnest, they should follow him to the Parliament Building. A tall, commanding figure with long, flowing hair, Perry led off in a buggy, the mob surging behind him through the narrow streets, denouncing Lord Elgin for signing the outrageous bill, but cheering the Queen.

The doors were locked as the sound of the mob approached. Soon their arrival was signaled by stones crashing through the windows. Presently, the main doors were battered in by the 35-foot ladder off the hook-and-ladder cart, and the leaders were inside.

A brisk encounter between the legislators and Perry's gang ensued, with flying inkstands, books and chairs added to the stones coming in through the windows. The mob gained control, the chamber was wrecked, the mace carried off and the members driven out. As the crowd left the chamber, according to Perry, he decided to smash the clock, still calmly ticking away above the ornate speaker's chair. On the third try, with a stone picked up from the floor, he accidentally struck a gas chandelier, breaking it. The flame ran up into the ceiling space, setting fire to the loft.

Other contemporary versions suggest the deliberate setting of fires at the other points. The Montreal *Transcript* said that the fire so quickly enveloped the building it must have been set in more than one place. This is plausible. The police and the army were conspicuous by their absence for a considerable period that evening.

The fire alarm sounded again and the firemen turned out their engines, but only to protect the surrounding buildings, including the nuns' hospital, Hotel Dieu. Not a drop was put on the blazing Parliament Building. The two Francophone engine companies, Voltigeurs and Hero, declined to turn out until promised safe conduct. Chief Engineer Perrigo, for some reason, stayed away altogether.

Fred Perry and several others were charged with treason and arson, but after three days of riotous partying in the jail and ugly mob scenes outside, the prisoners were released. No one was ever tried due to lack of preliminary evidence; the government breathed a sigh of relief when the riots, which included the stoning of Lord Elgin in his carriage, finally subsided. A month later the government moved to Toronto, quite possibly depriving Montreal of the chance of becoming the permanent capital of Canada.

Because the firemen and other prominent Tories had always declared their loyalty to the Queen, having claimed the government to be the traitors, they were quickly forgiven. Perry became a popular hero again and was even honoured by the government for his activity against the American annexationist movement and for service at the first two World's Fairs.

A Bantam Firefighter

Another prominent firefighter in mid-nineteenth century Montreal was William Orme McRobie, a volunteer who became one of the first career firefighters in the Montreal Fire Department, then captain of the Salvage Corps company, and it would seem, virtually second in command in the department. He should have succeeded Alexander Bertram as chief, but the ineffectual and antagonizing deputy, William Patton, took over. As the department slowly slipped downhill, McRobie resigned in disgust, writing a book of memoirs and advice as to how the fire department should be run. In 1882 he accepted the appointment of organizing the career fire department that took over from the volunteers in Winnipeg, and he was chief there until 1889.

McRobie emigrated from Perth, Scotland, in 1850, aged 13, and at 14 caught the North American enthusiasm for the volunteer fire service. He became an engine torchboy for the "Queen," Engine Company No. 5, on Wellington Street in a tough section of the Irish Griffintown district. In spite of his short stature, four feet six inches, he was quickly promoted to branch torchboy. This meant he went from lighting the way around the engine to accompanying the nozzle or branch men at the end of the hoseline.

The company held monthly drills in the spring. On his way to the engine house, dressed in his red Garibaldi-style fire coat, leather helmet, broad leather belt with the name "Queen" in large letters on it and his big iron hosekey hanging almost to the ground, the little fireman had to run the gauntlet of toughs on the street corners. On one occasion, after being seen to turn and challenge some of them to fight him one at a time, he arrived at the engine house to have a big fireman hoist him easily up onto the brake of the engine with the cry, "Now crow, little bantam, crow!"

At 16, McRobie was a regular fireman, two years before the legal age, proudly stepping up to receive the small quarterly stipend the volunteers were now being given. His career with the Queen soon came to an end, however, when his father had him indentured as an apprentice and he was not allowed to leave his job to run to fires. But no force can keep a confirmed firefighter away from fires, and young McRobie was eventually arrested for deserting his job. This did not work either; his father, his boss and the police could not frighten him into promising to give up running to fires. After a short time spent in Ottawa, he joined Montreal's No. 3 Company, "Protector," whose engine was known as "The Centre Ward Beauty," and he was back with his only real interest.

A Rash of Conflagrations

The burning of Parliament had emphasized the political liabilities of the volunteer fire department of that time in Montreal. Its inadequacy in controlling the severe fires that now could develop in the bigger buildings of the expanding city, where higher values on land created crowding, soon was dramatically illustrated. The occurrence of several fires (that would be the most devastating to occur in the history of the city before or since) proved the point. Also well proven was the fact that restriction of the construction of highly combustible cities was essential, as was the need to develop more powerful streams of water to fight well-established fires.

Volunteer Chief John Perrigo had at his command six engine companies, equipped with hand pumps capable of one modest stream of water each, and the Hook and Ladder and Hose Company. It was still the heyday of the city volunteer, with more enthusiasm and rivalry than science or resources. The waterworks system was still rudimentary and very limited for the area it covered. The technology for catching up was still just around the corner.

Fire chiefs in the last century had to be resourceful in the face of primitive tools. None was more inventive then Chief Decarie of Quebec, who had an ingenious nozzle and an early aerial ladder named for him.

On Sunday, June 7th, 1850, a fire started in a carpentry shop at St. Peter and St. Sacrement Streets, a hazardous premises in the congested old business district. It quickly spread, pushed by a fresh breeze, to St Andrew's Church, crossed St. Peter Street, an easy jump of only about 30 feet, then spread among the old stone buildings in defiance of Broomfield's idea that this construction would stop it. Poor maintenance of the old iron window shutters, gaps in the integrity of the walls, and the wind may have been the reasons. Inability to quickly develop a strong application of powerful streams undoubtedly contributed.

The congested commercial area bounded by St. Peter, St. Francois Xavier, St. Sacrement and St. Paul streets was cleaned out, leaving nothing but broken stone walls and a few chimneys. Notre Dame Cathedral and the Hotel Dieu, as well as the shipping in the port, were threatened as the fire reached Commissioner's Street then doubled back at St. Joseph (now St. Sulpice Street) to envelope both sides of St. Paul, the principal wholesale row. At this stage the fire department was helpless, with about 30 large buildings in three parallel lines having become roaring furnaces.

This conflagration became known as the Frothingham-Workman Fire after two leading merchants burned out, and was very costly for the insurance companies. When the firemen were about played out, Fred Perry organized the sailors off ships to make the final "stop."

The unusual condition, which insurance underwriters have always been aware of as a factor in conflagrations, was clearly there in a pair of larger conflagrations occurring two years later. The first started in the St. Lawrence suburb, up on the hillside running north of the old city. The first unusual condition was that the reservoir floating on top of the waterworks distribution system was empty due to repair work in progress on the lines. With a brisk southwest wind, unusual condition number two, a fire broke out in an outbuilding of Brown's Tavern at St. Catherine Street and St. Lawrence Main Street on the morning of July 9th, 1852.

Fanned by the wind, the flames rapidly ate their way through the wooden, lightly-built working-class homes of this outlying residential district, sending the inhabitants fleeing before the fast-moving waves of heat and flaming brands. The water supply was grossly deficient and sparks flew ahead of the flames, driven on the wind in a northeasterly direction.

The fire travelled along St. Catherine and Dorchester Streets to the open fields at St. Denis Street, destroying the magnificent new Bishop's Church and Palace. The General Hospital was saved, though fire raged all around it. Fred Perry, not in the fire department at this time, received the formal thanks of the hospital governors for his help in rallying the firefighters to save the buildings. The insatiable flames burned their way down to Lagauchetiere and Elizabeth streets, and almost south to Craig Street before running out of fuel, having destroyed the cattle market and other buildings on Viger Square. Over 600 houses were burned, mostly working-class homes where there would be little insurance.

This had been a major conflagration, destroying the northeasterly ward of the city. The exhausted firemen were beginning to ease up on their exertions by about 5 p.m. and to enter the mopping up stage, only to hear the peal of alarm from the church bells once again. Sparks from the big fire of the afternoon had carried into the east end of the business district to start an unseen fire in a building on Dalhousie Square. Very quickly the flames caught up the flammable back sheds and fences behind the important and impressive buildings facing on the square. The Hayes House, the leading hotel, and the Theatre Royale were soon engulfed, followed by the army messes and the homes of the aristocracy. Champ de Mars Street, and St. Louis and Durham terraces succumbed next.

By late evening the fine old square was gone and the fire running, without much hindrance from the fire department, northeast along what is now Notre Dame Street East, and along the waterfront streets between it and the harbour. Firemen and citizens fought it in vain as it progressed as far as Gain Street in the east. A smaller fire in area than the first, this evening conflagration destroyed several churches, a brewery, various industrial establishments and many fine cut-stone houses of the upper class.

Yet a third fire, also caused by flying brands from the original, burned out sawmills and

Hayes House hotel and the other grand edifices of Dalhousie Square go down on the disastrous day of July 9, 1852, in Montreal.

lumberyards on the waterfront in the east end of the city. Fred Perry was still trying to organize firefighting operations after 24 hours of labour, and was given lavish praise by the Montreal *Gazette* for his efforts. Not so Fire Chief Perrigo, who was severely criticized and soon after resigned.

Following this day of calamity, in which an estimated 10,000 inhabitants out of a population of 57,000 were made homeless, tougher regulations were enforced on wooden buildings, so that the rebuilt St. Lawrence suburb became a better-class district with many fine homes. Montreal suffered a renewed state of depression in the fifties for reasons of trade, but the setback of having a large part of the city destroyed must have contributed, too. As in most cases, there is little concrete data to indicate the actual effect of such a fire on the economic well-being of the city. Adding to the burden of the losses was the burning of the two cathedrals, Roman Catholic in 1856 and Anglican in 1857.

When the discredited Chief Engineer Perrigo resigned, he was succeeded on September 13,1852, by Alexander Bertram, 41, an easygoing blacksmith-machinist, veteran volunteer fireman and one of the two assistant engineers under Perrigo (the other was Captain John Fletcher, who had made the great run to Laprairie). Remaining as chief until 1879, Bertram bridged the transition from volunteer to full-time professional fire department. He travelled widely and was popular with his men. Unfortunately, in his last years, the oldest fire chief in North America, his control of the department slipped and his firefighting operations were criticized.

Bertram managed to keep the volunteers under control and to keep their minds on firefighting, most of the time at least. There were dust-ups between the French Voltigeurs and the English companies, but the greatest problem was the continuation of the battle between the Orange and the Green, the companies supporting their ethnic district loyalties. The Griffintown Irish were notably fractious. It is said that at a serious fire in the Griffintown quarter the Orange connected companies were stoned and driven off by the locals.

By 1858, when the large numbers of volunteers were becoming redundant, they became more militant and jealous of every prerogative. The Union Company was very obstreperous, and Chief Bertram finally recommended its disbanding. Bertram won and the Union was folded up.

There was very little spit-and-polish in the early years of the professional fire brigade as shown by this motley crew at a Montreal station. Each man dressed as he wanted or could afford.

A Waterworks System Brings Changes

After tentative starts in 1816 and 1819, and even earlier as a private company, the Montreal waterworks system became of some value in firefighting by 1832. It utilized four-inch-diameter pipes and a steam pump on the St. Lawrence riverbank, with balancing tanks in a building on Notre Dame Street, which gave a head of a mere 27 feet on that street. Water carriers were also still used and the hand fire engines employed to boost the water system pressure to supply the hose.

In these earliest days, water for firefighting was obtained by removing a plug in the waterpipe at the bottom of a pit in the street. Hence the name fireplug, sometimes still used for hydrants today. It was a slow and cumbersome method, soon replaced with the familiar post hydrant. New pumps, a reservoir and mains extending into the Griffintown area to the southwest were added in 1847 and 1849.

The ambitious Lachine Aqueduct project was begun in 1852 to bring water from the Lachine Rapids down to Atwater Avenue on the western edge of the city, a distance of nine miles, by way of a nine-foot-diameter conduit. The plan was to utilize the pressure of the water at the lower end to drive pumps supplying the city distribution system. The hydraulic power did not prove sufficient and steam power had to be installed for the pumping. At any rate, the water was now much cleaner than the harbour supply.

The new water supply system was completed and opened in 1856 with unheard of pressures now available, capable of providing strong hosestreams directly from the hydrants without the need for supplementary pumping by the fire department. Even so, it took two years before the fire company captains were ordered to reduce their membership from 40 men each to 12. The companies would now run with their hose jumpers (or reels) only, no longer needing to haul or operate the heavier engines. At first they had insisted on running hose from the hydrants into the tubs of their engines and pumping it by hand to the fire, in a fine example of early featherbedding.

The manpower reduction was the first move toward the disbanding of the large force of about 400 volunteers and the end of a useful but trying element in the community. There had been agitation for a full-time professional fire department during the fifties, as soon as the higher-pressure water supply system permitted. Chief Bertram favoured it, and Fred Perry suffered a few bruises at the hands of his erstwhile friends for advocating it as well. The great attribute, in the view of these two veteran firemen, would be a faster initial attack on fires.

In the autumn of 1859, the city decided to employ one full-time "guardian" in each of the nine stations who would maintain the station, see to the sounding of the alarm bell when people came to report a fire, and respond with the horse-drawn hosereel. On April 24, 1860, they went on duty.

With this preliminary step, the many volunteers who were nearing a length of service that would give them permanent public service exemptions were being accommodated. They would be phased out with their exemptions earned and the professional or career force increased during the next seven years. But the heyday of the city volunteers was already over. It was a powerful hydrant system that spelled their end in Montreal, not the steam fire engine as elsewhere. The speed and simplicity of direct hosestreams from hydrants that could deliver over 400 gallons of water per minute each made the Montreal Fire Department a model of effective firefighting for some time.

In this mid-century period — with its several periods of economic recession — in Montreal there were said to have been many arson fires to collect insurance, or "selling to the English," as it was put then. The insurance companies were very tough, agents investigating fires and often beating down or refusing suspicious claims.

A great fire tragedy occurred on the St. Lawrence River in 1857 when the steamer *Montreal* caught fire about one hour out of Quebec, on the way to Montreal with 500 newly arrived immigrants. There was a panic and the captain ran his ship onto the rocky shore, but many lives were lost, some 250 bodies being recovered after the fire. The steamer *Napoleon* later arrived in Montreal with a few survivors and the news.

(XII) A Mix of Firefighters

The On-Call Fire Companies

While the reorganization of the old totally volunteer fire companies into the on-call type reduced their size and ability to throw their weight about, it did not always eliminate their competitive or social spirits. Now they were smaller in numbers, paid a token for their services and generally began to arrive second on the fire scene to aid the skeleton force of full-time, career firefighters.

The career members, full-time on the city payroll, were generally only steam engineers, chiefs, drivers and stokers for another 50 years or more in all but the largest cities.

Nevertheless, they pitched in and performed a lot of actual firefighting before the on-call men arrived. There are many small fire departments so manned today, the few career men handling many small fires so sparing the volunteers the nuisance and time lost from other jobs that response to these would entail.

But the parading in dress uniform finery, the proud mottoes, the parties and competitive spirit were, in many cities, not so very much diminished. The new on-call firefighters, whether paid or not, were still called volunteers and were keen firefighters all. Volunteer companies still had their favourite politicians and received private donations towards the comforts of their clubrooms. Just as today, there was a blurred distinction between the wholly volunteer fireman and the on-call paid man who received a small stipend or per-fire sum. Generally, both are called volunteers today.

In Saint John and Halifax, the new companies of "call firemen" soon developed their own identifications and mottoes. They also competed on the sports field in "musters" with the still-existing all-volunteer companies from smaller centres. But being paid, however little, they were servants of the municipal corporation, which made a subtle difference.

The coming of the steam fire engine and the additional hose-carrying response needed to exploit its great pumping capacity also dictated the use of horses to pull the increasingly heavy equipment. This, in turn, was another reason for men on duty around the clock in the stations — to look after the horses. This was especially so in the hilly cities like Saint John and Halifax.

The first career, or professional, firefighters in Saint John were engine and hosecart drivers, along with the engineers who operated the steamers. Gradually, though, as the fireload increased, the career force expanded. By the mid-1870s, under Thomas Marter as chief engineer, the Saint John department direction consisted of three on-call district chief engineers, each in charge of a sector of the city. Each of the three engine and hose companies had an engineman, an assistant engineman who drove the steamer, and a hosecart driver. The ladder truck also had a driver.

Halifax was the first city to challenge the volunteers' power by installing steam fire engines. Here is "Albert", changed from hand draught to horses because of the city's steep hills.

Typically all these career men lived continuously in the fire stations, with time off to go home to meals and an occasional day off. Even with this skeleton force, six to eight men would arrive with the apparatus, enough to commence an attack on the fire before the arrival of call men, who came on foot. During meal hours this force was halved, and those at home had to drop their forks and run when they heard the tower bells strike an alarm.

In Saint John in the 1870s there were altogether 35 on-call part-paid men at $30 per year in the three hose and one ladder companies. A coal wagon driver was on-call, and there was an electrician to maintain the fire alarm telegraph box system. The old volunteer fire police company had dwindled away to nothing by this time, and the new Salvage Corps and Fire Police had not yet been organized. There was some criticism of the weakness of the fire department and its organization under an aging former volunteer, and this became vociferous following the great conflagration of 1877, notably as to the city's policy of using the firemen and horses to do road maintenance work far from their stations.

Saint John continued with a modest career force backed by the call firemen, gradually increasing the former while the latter dwindled away around the beginning of World War II. Two all-volunteer salvage corps and fire police companies, organized in 1882 and 1898, continued on, however, and combined as one company in 1982. They still serve in the unpaid old tradition, alongside the modern career force.

Alderman John Kerr was a patron of the new No. 2 Hose Company, as well as chairman of the Fire Committee of the Common Council in 1877. An exchange of correspondence in the press early in the year perhaps illustrates something of the loyalties and sensitivities involved. It was common for a business firm suffering a fire to take an advertisement in a newspaper to thank the firemen for their work and make a donation to their funds. John Vassie and Company of Saint John wrote in an ad a message addressed to Chairman Kerr, acknowledging the valuable services of No. 2 Company in saving their property on the night of March 8th. They also donated $100 to the company's funds. A retort very quickly appeared as a letter to the editor over the nom de plume "Fair Play," objecting that the chairman's pet fire company was being unfairly favoured. An argument was given on the relative positions of Nos. 1 and 2 Companies in fighting the fire. It was claimed that No. 1 had been the key performer.

Next came a letter from Andrew Lawson, foreman of No. 2, to rebut the arguments of Fair Play and offer No. 1 half of the money. No. 2, Lawson said, was only interested in the good opinion of the Vassie Company on their performance. All this impelled the poor Vassie Company to protest that they had acted according to their own assessment and to absolve Alderman Kerr of any responsibility in their act of appreciation.

On the social side, No. 2 Company scheduled a grand sleigh ride the following January 8th, (The greater part of the city had been wiped out in a single conflagration in the interim, but life went on.) As so often happens in coastal towns, the snow had melted completely away by the scheduled evening. As an alternative, a parade with their hosewagon and the 62nd Regiment Band, hired for the occasion, moved through town after a band concert at their fine cut-stone engine room house on Sydney Street, King Square. This building is, incidentally, a rare case of a fire station built for hand fire engines that is still in existence today, having served as a fire station for over 120 years.

That evening, the official party of the fire company was followed about by a large and delighted crowd. The company secretary performed as parade marshal, leading the group to Alderman Kerr's house on Queen Street, where Mrs. Kerr plied them with refreshments for a half an hour. They drank to the popular Mr. Kerr's health, and the company sang "For He's a Jolly Good Fellow." To spread their goodwill, the group marched across town to Alderman Duffell's house and to that of Mayor A. Chipman Smith on Cliff Street for repeat performances. The parade then repaired to the engine house "for a further social time."

Occasionally in the Maritimes a great tournament, or muster, was scheduled. The visit of Governor General Lord Aberdeen to Saint John on August 14 and 15, 1893, was ample excuse. A number of volunteer and call fire companies from New Brunswick, Nova Scotia

Members of Dartmouth, N.S., Axe Company, predecessor of the ladder company, with their primitive ladder cart.

Interdepartmental musters were great occasions for both volunteers and call firemen. Yarmouth's Neptune Company are shown ready to depart for a visit to Saint John, N.B., in 1893.

By the 1870s, Saint John was well equipped with steam fire engines. This Amoskeag two-cylinder tank engine and its four wheeled hosereel fought the great conflagration of 1877, one flank of the fire being turned here at their engine house.

and Maine took part. The Saint John companies included Wellington No. 1, Union No. 2, Extinguisher No. 3, Prince of Wales from Carleton, No. 1 Hook and Ladder, and Aberdeen Hook and Ladder No. 3. Added to their rigs in the events were carts and reels from Fairville, Moncton, Amherst, Sussex and Woodstock.

Naturally there was the ever-popular evening parade by torchlight, with no less than seven brass bands. A daylight parade around the central part of the city and past the Governor General's reviewing stand at Sir Leonard Tilley's residence was more formal. On the second day, the races and competitions were held for a championship purse of $250. Thousands of spectators attended the events. Amherst won the title in the ten-man hosecart race by running 900 feet and laying out a 200-foot line of hose to produce water in one minute flat. Saint John took first for the ladder raise. To do this they ran 50 yards with a 30-foot ladder, raised it and got a man to the top, all in 17.4 seconds. Standard competition "evolutions" much like these are still features of volunteer firemen's musters.

Debacle in Saint John

The year 1877 was a bad one on the fire front for New Brunswick, It started on March 7th with another of those major business district fires at the foot of the King Street hill in Saint John, an area ravaged several times before, flames erupting among the business buildings on Prince William Street. Several major commercial houses were burned out, with a loss of $2,000,000, and a collapse took several lives. As they were brick replacements following a previous fire, there were brick division walls to help the fire department gain control. But there was an altercation over the actions of some of the chief officers that suggests all was not harmonious in that organization.

The chief, Thomas Marter, was not on hand and the steamers did not respond on the first alarm for some reason. George Drake was in charge as on-call district chief, but Andrew Lawson, foreman of No. 2 Hose Company, was there and was the firemen's fireman and natural leader. Lawson acted on his own initiative without agreement from Drake. As a grand jury cautiously commented later, it was the public perception that there were inadequacies in the leadership.

Also this same Spring there were major conflagrations in the business districts of St. Stephen and Woodstock, causing heavy destruction, on May 14 and 17 respectively. Five days later, Portland had a serious blaze at which Saint John helped out. The St. Stephen fire destroyed 150 buildings in the business district along the St. Croix River, lack of access on the water side apparently hampering the firefighters. The engine from Calais, Maine, was instrumental in halting the fire's progress, it having travelled a quarter-mile along Water Street in half an hour. Two steamers were trapped and saved with difficulty. Boatloads of goods were reported to have been stolen and spirited across to the American side of the river during the excitement, without benefit of customs inspection.

Saint John, N.B., and St. John's, Nfld., the namesake cities with dismal records of big fires, suffered what are likely the two most staggering conflagrations to occur in Canada's major cities, the former in 1877 and the latter in 1892. At the time, Saint John was one of the country's leading seaports and manufacturing centres, just beginning to pass its peak because of the decline of wooden ships and the post-confederation centralization of industry. This fire was a severe blow to the commercial heart of the old Loyalist city of 28,800. The city was rebuilt, but then entered a long period of little growth and gradual loss of competitive position. With many other factors involved, it is difficult to assess the role of the fire in this decline.

It was on Wednesday, June 20, 1877, a day that dawned cloudless and with a promise of more of the exceptional heat that the city had basked in for six solid weeks. Not a drop of rain had fallen for a month and temperatures had been 75 to 80 degrees for some days. A gusty wind blew up clouds of dust from the streets as the day progressed. It was very unlike Saint John, its weather normally variegated, with moderate temperatures, cool rain and fogs

The great fire that destroyed the heart of Saint John, N.B., in 1877, probably started from a spark from the chimney at right which landed on the shed just above the bow of the schooner. High winds carried the flames through the crowded buildings beyond.

A contemporary painting portrays, without much accuracy of detail, the conflagration sweeping across the peninsula of Saint John.

interspersed with clear days, always affected by the conjunction of warm land air with the cold waters of the Bay of Fundy. The unusual dryness had evidently resulted in some fires in the woods to the north and west, and smoke from them was plain to be seen, delaying the realization by many that a major fire was burning in the city, even after the fire alarms sounded.

Getting a start on a long buggy trip early in the morning was Thomas Marter, the aging fire chief, accompanied by a member of the city fire committee — they were heading upriver to buy several new horses for the fire department. They would not be back until after 6 p.m. As the day got under way, industry came to life with the sound of the caulking hammers from the shipyards in Portland and the whine of the big sawmill of Kirk and Daniel at the very head of the harbour on the west side of the North Slip, where the head of Long Wharf is now.

On the east side of the slip was York Point, one of the earliest built-up parts of Saint John, and by then a very crowded tenement district jammed with two-and three-storey houses. On the harbour side were small wharves occupied by ship chandlers and storage sheds. There was a string of them running down North Wharf along the North Slip. These were also known as Fairweather's Wharf and York Point Slip. The location is to the north side of the western extremity of Union Street, beyond Mill Street.

In the early afternoon, about 2:30, a young coalheaver, Dennis Costigan, looked up from his heavy work on the brigantine *Belle Waters* and across the slip. His eye caught a small flame on the roof of Henry Fairweather's storage shed on the opposite wharf. He told his fellow workers and ran around to the other side of the slip. "The flame was about a foot and a half in width on the roof," he said, "I ran down the alley and got on the roof. I tore up a board and tried to make a hole in the roof to put the fire out. Just then I heard the York Point fire bell ringing. I think I was on the roof about a quarter of an hour. I stayed until the hose came. I did not get the fire out. It ran up the roof and I got down. The hosecart came and they put the hose inside the barn."

So it started. Afterward there was much debate over whether the fire started inside the large open loading doors of the shed's second-storey loft, full of drying hay, or on the roof. It was generally felt that it was a spark from Kirk and Daniel's mill chimney stack across the slip that started the 13 hours of destruction, but it could have been spontaneous ignition in the damp hay. Costigan said it started on the roof.

Fairweather's shed measured 40 by 80 feet, an old wooden storey-and-a-half building squeezed between two others on the wharf and containing salt and hay. One of Costigan's mates ran up the wharf to ring the York Point fire bell on Smythe Street, still used, but a holdover from a decade before when the electric telegraph boxes were installed. As customary, when he heard the bell, William Lantalum, a nearby junkdealer and volunteer fireman possessing a box key, ran over and tripped alarm box No. 5 at Union and Dock streets.

At the time, the Saint John Fire Department had a career force consisting of the chief, an engineer, engineman and driver for each of three Amoskeag steam fire engines (with a fourth as a spare), three hosecart or reel drivers, and a hook-and-ladder truck driver. An on-call force included three district chiefs, a captain of hook-and-ladder, three hose company foremen and 38 firemen. The Carleton district of the city, west of the harbour, had an engine and hose company, apparently volunteer and operated pretty much independently of the city department. Portland, adjoining Saint John on the north, had a hose and engine company and some hosereel houses, as well as hand-engines in reserve.

The three engine stations and one ladder station in the city proper all responded to the box alarm. There was undoubtedly some realization of the hazards of the city since the entire force was scheduled to turn out on the first alarm. Following city policy, but with a ghastly error in organization, the horses of two of the three steamers, with their drivers, were spending the day at the opposite end of the city on road maintenance work.

District Chief Samuel Dunlop was the first fireman on the scene. He was on-call chief of the district and had been in his ship chandlery office two doors down the wharf from Fairweather's. No. 3 Hosereel driver Jackson arrived and promptly laid a line of two-and-a-half-inch hose from the hydrant at Smythe and North Streets, down the wharf to where the flames were now billowing from the loft and roof of Henry Fairweather's shed, blown by the north wind entering the big doorway facing the slip.

The steamer was connected and a good strong stream of water directed into the building. But on the dangerous downwind side, No. 2 Hose arrived and was only able to put a line in operation under the lower direct hydrant pressure. Here a neighbouring shed was already ablaze. No. 1 Steamer, arriving 15 to 20 minutes later, took a position at Mill and Pond Streets, far east of the fire and had to be moved closer, to Union and Smythe, south of the fire, so as to better head it off.

But the delay had been too great and it was too late to work here directly downwind of a fire that was now racing through the flimsy, crowded sheds on York Point Wharf and raining sparks into Smythe Street. Speed of intervention was of the essence, but the career force had been delayed, the steamers late, bystanders reluctant to help, and the call force trickled upon the scene slowly.

The old frame buildings burned like kindling, the heat chasing No. 2's hoseline out of Ramcat Alley, a critical line at the head of the Wharf, showering flaming brands onto the tenements of Smythe Street, two blocks of ancient lower-class homes, all wood. In front of the fire the weaker forces were retreating from the wind-driven onslaught, but there was no change in firefighting tactics forthcoming as the flames marched rapidly through the York Point neighbourhood toward the main thoroughfares of Mill and Union Streets, the principal wholesale and business districts lying beyond them.

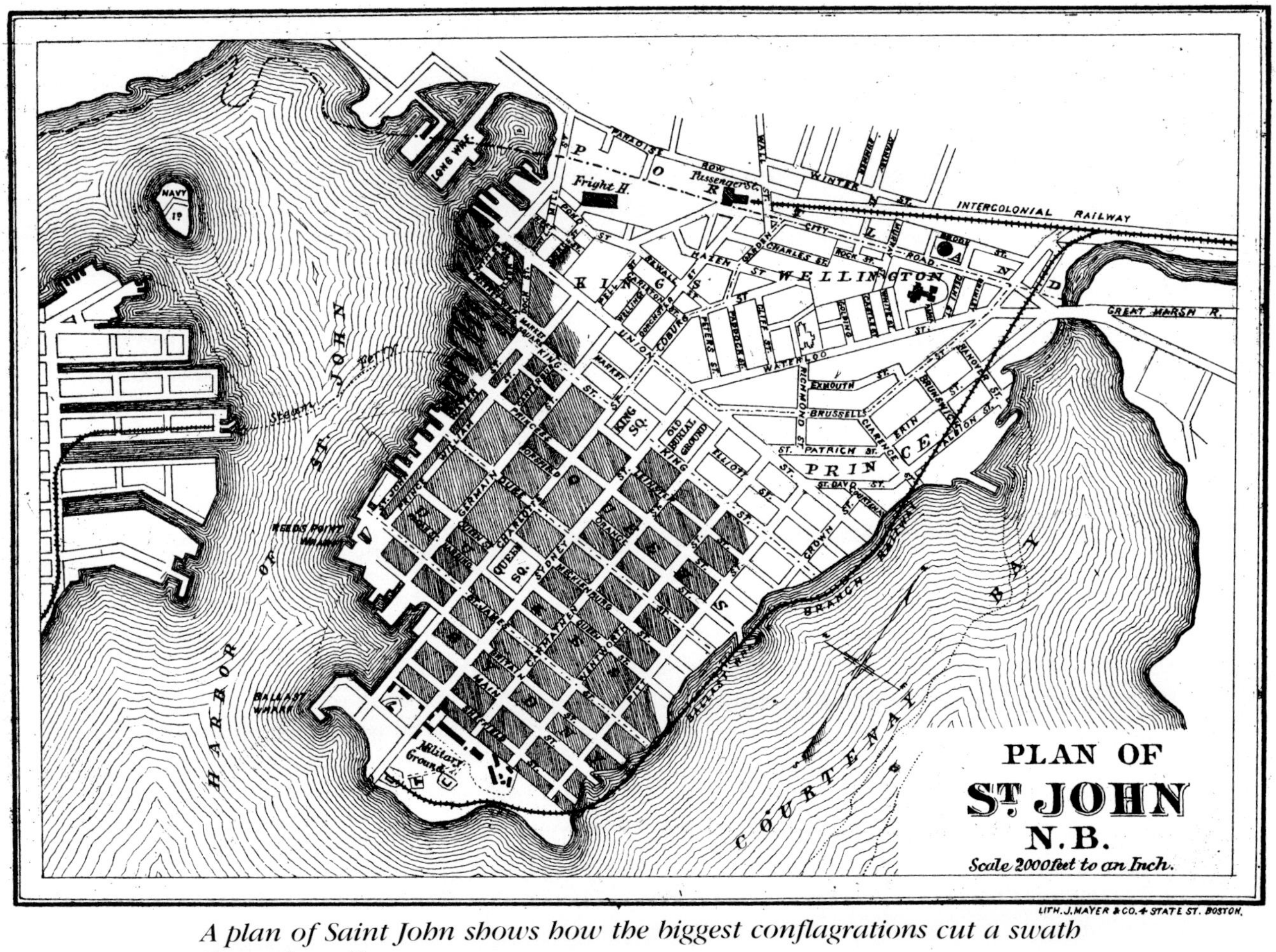

A plan of Saint John shows how the biggest conflagrations cut a swath through a city.

Gobbling up everything in its path, the fire leapt across a narrow slip to Hare's Wharf, and thence by another slip into the wholesale warehousing area of North Wharf by way of the waterfront. This is where the present Market Square shopping complex stands. Further back from the waterfront a momentary gust carried solid flame right across Union Street and the brief touch was fatal. Firemen held boards before their faces and held their ground until their clothes were singed, then were forced to drop their hose and run, then drag their hose out from a safe distance up the hill. The crossing of this street by the fire eliminated a good potential defense line, but there had been no coordinated attempt to establish one.

From the formal enquiry afterward, it is clear that Dunlop had no conception of the need to head off the swift-moving fire, either by getting his forces in front of it or, failing this, trying to pinch it off with attacks of the flanks. He could only see the fire in three or four buildings in front of him and stubbornly insisted on committing his forces, the entire department for a time, at that one point. He was actually chasing the fire from behind, from the windward side.

Embers flying very high on the gale, augmented by the thermal blast of the conflagration, soared far into the business district of Prince William Street and apparently quite early set fire to an old building on the military grounds. This was in the extreme south end of the city. There were no firemen to answer a call if one was even made. The flames of this fire quickly spread into the houses on Main Street (now Broad) and across its great width to set the second conflagration going amid the industries of this Lower Cove district, including a sewing machine factory, foundry and lumber yard. The sight of this fire came as a shock to the throngs making their way south along the waterfront streets away from the original fire, eventually trapping some of them on the wharves.

The firemen moved from defeat to defeat, a fragmented and bedraggled army, driven back constantly, dragging a few precious lengths of hose and a nozzle, in most cases working without benefit of a steamer to boost the hydrant pressures. These were steadily weakening as abandoned hoselines and broken pipes in destroyed buildings bled off the otherwise strong pressures of the gravity waterworks system.

The main fire swept across the wide Market Square via the vessels in the slip without pause and through what had been thought to be massive fire-resistant mercantile edifices around the square by attacking them from behind. Onward it went, down Prince William and Saint John (now Water) Streets, wiping out the monumental stone banks, the beautiful new post office, the elaborate customs house and the lowly wharves and sheds on the water side.

As the afternoon wore on, the wind came around more from the west, turning the fire uphill from these streets into the very centre of the peninsula on which the city was built. At 5 p.m. the fire was well spread through the city, except for the eastern half. There were many solid brick-walled buildings, but many of these had combustible roofs clad in wood shingles, and there were many wood-frame ones among them as well. Working more easterly now, the fire doubled back through Canterbury, Germain and Charlotte Streets to the top of the hill, joining other fires to cross the peninsula down to the water's edge at Courtenay Bay by the early hours of the morning.

In the light of morning, where a bustling and prosperous city had stood, there was a scene of desolation; two-thirds of it was gone. The monumental buildings and humble tenements had vanished, leaving a forest of chimneys and smoking rubble. A city built by entrepreneurs and working men to a position of prominence had suffered a great blow, just at a critical time, for Saint John was beginning to feel the pinch of more difficult times to come.

There had been 1,612 buildings destroyed, with a loss of $27,000,000. About 13,000 people were homeless. The major national and international insurance companies paid up promptly and even made donations to the relief fund in addition, but the small companies and mutuals with heavy local liability had difficulty, several going up in the smoke without so much as a stick of furniture left of them.

The treasures and heirlooms of the Loyalist descendants, the palatial homes and private libraries of the wealthy shipowners and lumber kings were gone, but the human reaction was action to rebuild immediately what they had lost. This time Saint John rebuilt in brick and stone but only within the business and commercial areas. The residential blocks again became a solid mass of wood.

Many of the homeless from this great fire were thrown on the streets again in October, 700 of them, along with 1,300 Portlanders, being burned out on the 20th as a fast-developing outbreak destroyed several blocks along Portland Street, from Main Street to the harbour. The Saint John Fire Department had the opportunity to return a past favour by rushing to assist.

There was much acrimonious recrimination following the Saint John disaster of 1877. The fire occurred under conditions of very dry, windy weather, massed wooden buildings, wood-shingle roofs and a fire department without competent direction. The water supply was good until overwhelmed by wastage. Individual fire officers worked heroically and intelligently, but there were obvious failings.

Fire Department Without a Chief

Halifax differed from the norm in these times, operating for a long period under a Committee of Management of the Board of Fire Wards. Without an operational chief of department, the chairman of the board or the first board member on the scene of a fire was in command, a system virtually unchanged over a century. Instead of a number of fiercely independent fire engine companies, the equivalent was set up as divisions of a single Union Fire Engine Company. This seems to have stifled any tendency toward strong rivalries. On the replacement by a new one and the reconstitution of the old company as the Union Protection Company, Halifax continued to be a city in which the number of citizens involved in firefighting was very large.

The U.E.C. of 1860 had six divisions, named Rapid, Vesta (formerly Alliance), Alert (formerly Halifax), Resolute, Aetna and Alma. The Axe Company was the Halifax equivalent of a Hook and Ladder Company. The company hosereels often had their own names, too, like Perseverance, Salamander and Dreadnought. Independent now of the city fathers, the 80-man protection company eventually had its own three-storey building with a stable, wagon house and social rooms on Jacob Street. It received a one thousand dollar grant from the city, however, and maintained a pair of greys with driver to pull its wagon of salvage bags, baskets and covers.

The Maritimes-New England connections were very strong, and Haligonians were obviously aware of the latest in firefighting developments to the south, so Mayor Caldwell set off at the end of 1860 to look into and then buy Canada's first steam fire engines. These were two early-model Amoskeag engines built by the Manchester Locomotive Works in Manchester, New Hampshire. Of hand-drawn type, they were loyally christened Victoria and Albert. As elsewhere, this move immediately diminished the grip of the volunteers on the city.

Still, by the mid-seventies, the fire department had the same six engine houses and the fires were still fought by the same volunteers. But manhandling the 3,500-pound steamers over the steep hills of Halifax proved impractical, to say the least, and they were converted to horse draught. Stalls were erected in the engine houses and five full-time drivers engaged. Nine horses were used under contract for some time, but replaced by city-owned animals in 1884 and 1885.

Rubber hose of the Gutta Percha patent was replacing the leather variety rapidly in the seventies and eighties, no doubt to the relief of the firemen. The maintenance of the riveted leather hose was both critical and laborious to avoid its becoming as stiff as a board. In 1879 Halifax had 3,000 feet of rubber and 5,500 feet of leather hose, proportionally about average for the major cities.

Probably a family picture, proud members of St. George Company No. 1 of Yarmouth, N.S.

Halifax was one of the many cities that utilized career and on-call firefighters together for a long period of time. Here are the call men of the Spring Garden Road station with their Cole Bros. steamer and a hand cart brought out for a parade.

Dissatisfaction with the fire department organization in Halifax festered in the latter 1880s. A test fire was staged in the Public Gardens in 1890. The results were seemingly controversial and editorial comment suggested that a paid fire chief, not an amateur board of fire wardens, was needed. The alderman in charge of the fire was described as a better tailor than fire chief.

It was not until 1894 that the city approved a plan to replace the strictly volunteer force with a paid department, that is, an establishment partly of career men and the balance a call force for whom it would be a part-time job. The new fire department included the first chief, John Connolly, paid $800 a year, a fire alarm electrician, two chemical engine operators, four steamer enginemen, 16 drivers and 14 hose-and-ladder men, at $416 each, along with 62 call firemen paid $62 per year. The total budget the city was now committed to was $27,274 a year.

After 1900 there was, as in Saint John, a growing career force and declining call force. In 1919, following World War I, a fully career force was formed, though the Union Protection Company continued in its salvage capacity for a little longer. The department was reequipped for the combined career-call force during 1892 and 1893 with the addition of a fifth steamer, the first chemical engine, Acadia, and eight hosewagons to replace the two-wheeled cart reels, as was now being done elsewhere.

The Police Take Over in St. John's

Accounts of the volunteer fire companies existing in St. John's Newfoundland, from 1846 to their end in 1892 are sparse. References to the Phoenix Fire Office forming "an important fire company" after the fire 1846, with an engine house located at the foot of Long's Hill, are recorded. It is possible that this insurance company made a special effort following its continued losses in the periodic conflagrations in the old port city and the company of firemen may have been paid for their work. The company was disbanded in 1876.

There are mentions in accounts of major fires of "two regular" companies, as well as military and private engines. These accounts record, "The Phoenix and military engines were ineffectual" at a minor conflagration in July 1855; "The Phoenix and Cathedral Brigades responded immediately to the call of the bell" in 1859; and "Charles Fox Bennet and his engine rendered excellent service." One account referred to "the Phoenix and Volunteer Companies," as though there were a different status, such as pay for the former.

The impression is that the Phoenix, Cathedral and the army garrison were the only continuing fire companies, and that they were not coordinated as a single fire department nor under a single chief. This would have been the situation until 1867, when the companies were disbanded on the formation of a St. John's Volunteer Fire Brigade on July 4, 1877.

During the life of this new brigade the key differences must have been organization as a single fire department and government involvement. The first steam fire engine went into service in this period, a gift of the Phoenix Assurance Company, while a second one was the first government financial contribution to the fire protection of St. John's.

St. John's had continued its notorious record for great fires into the second half of the nineteenth century. Big fires occurred in July 1855 and two more the following year, with 90 houses lost in one of them and the other in the higher levels of town where water was scarce. On the whole, though, there were optimistic reports that the fire companies were now doing good work and stopping many threatening outbreaks.

The perennial fire nemesis of St. John's was bound to have one last great binge, however, before tougher building laws and a strong firefighting capability gained control. It came in 1892, with ideal conflagration weather,a rarity in St. John's humid climate. The summer had been hot and dry, and July 8 saw the city smoky from brush fires in the outskirts. It was 87 degrees Fahrenheit, very unusual for this coastal city. Brisk winds blew dust along the streets. It was like Saint John on its day of disaster 17 years before. True to

The last and greatest of the numerous conflagrations that swept St. John's, Newfoundland, occurred in 1893, leaving this scene of desolation. Major countermeasures followed this one. - PAC-PA66621

the insurance underwriters' code, in which the unusual event triggers the conflagration, there was a special problem for the fire brigade.

The waterworks system had been shut off for extension work and, though turned on again in the afternoon, had not yet filled the pipes to the upper levels. Also an old fire cistern at the top of the hill had recently been emptied by the firemen in a drill and not refilled.

Given these conditions, along about 5 p.m. a teamster committed the simple careless act of dropping a match or his lighted pipe into the hay in a barn or stable on the heights of town at Freshwater and Pennywell roads. The hay-filled building rapidly became totally engulfed in flames, and these were spreading to other adjacent structures under the urging of the wind gusts when the fire brigade arrived.

With no water for firefighting, the flames leapt from house to house unhindered. Down Carter's Hill, Harvey Road and Long's Hill into the centre of town, by nightfall the roaring inferno was in the heart of the business district along the waterfront.

Once more the great business houses of the Water Street barons were consumed, wasting the years of entrepreneurship and fortune-building. Also lost in the sweep of the central body of the city were the principal government buildings (other than Colonial House), the Anglican Cathedral and the Atheneum Hall. Some of the vessels at the piers behind the waterfront warehouses were unable to get under way in time and burned.

The first fire headquarters at Fort Townshend in St. John's Nfld., a substantial fire hazard in itself.

Steam fire engines and putting firefighting in the hands of the Newfoundland Constabulary were part of the solution chosen by the government after the fire of 1893. The police-firefighters in English brass helmets ride their Merryweather engine.

Brushfires were set on the distant South Side Hills as embers set new blazes going south and east, well ahead of the advancing flame front. Even the brick and stone buildings that had now become fairly common in the business district succumbed in the blizzard of flame and embers from the wooden housing adjoining them.

By morning, the conflagration had cut a wide swath through the greater part of the so frequently devastated old seaport city, as far east as Signal Hill, the boundary in this direction below the old railway station and the eastern extremity of the city. Everything was destroyed below Harvey Road and Military Road, from Carter's Hill eastward. Two-thirds of the city, 2,000 homes, were smoking ruins in the light of dawn. As is frequently the case in this kind of conflagration, when there is a great hubbub and most people are not in bed, there was plenty of warning for occupants to escape. Only two or three lives were lost, according to the accounts.

The monetary loss was set at 20 million dollars, but it is not possible to assess the loss to ongoing commerce and prosperity following such catastrophes. Loss of regular markets and jobs are offset in part by reconstruction activity, as insurance claims money pours in to cover part of the loss, but in this case it was only about $4,500,000. As the persistence of life is the most striking feature in the biological world, so too seems the persistence of human endeavour in commerce in the face of calamities like fire. Water Street was soon rebuilt.

Unfortunately, man has always been slow realizing that the prevention of fire catastrophes lies largely in his own hands. But this time in St. John's they had had enough. The rebuilding of Water Street was in solid brick and stone, and wider. Regrettably, the many other streets behind the business district must have been considered expendable on into the future, for the commercial core was again surrounded by wood-frame ranges.

A prominent step against the depredations came as the result of an enquiry into the great fire. It was critical of the fire brigade's weak manning and inadequate equipment, even though some good effect from the firemen's efforts was recorded. The government decided to replace the volunteers with a career fire department. Following British concepts, the job was turned over to the Newfoundland Constabulary. The immediate solution of Inspector General of Police McCowan was to barrack his single constables in the three fire stations that were built. They would provide the first response to fires. He thus had 19 police-firemen, or fire constables, and 102 others as reserves.

Three fire stations were erected: Central, opposite St. Bon's Gate; the West End, at New Gower Street and Bambrick; and the East End, at the foot of Ordnance Street. A harbour tug was equipped and retained as a fireboat. St. John's being the essential life-supporting hub of the island, the Newfoundland government footed the bill for all this. The St. John's Fire Department is still operated on this unique basis in the 1980s. It remained a division of the constabulary until the late 1950s, under a fire superintendent reporting to the chief of police, and with the firefighters classed as fire constables.

By the end of 1895 each fire station had a British-built steam fire engine, North American-style hosecart and horses to haul them. A ladder truck was added the next year. The firemen joined in annual parades with the constabulary and were prominent in the celebration of Queen Victoria's jubilee in 1897 with torchlight parades and illuminations.

The fire stations were large wooden buildings, four storeys high, providing living quarters for the district chief and his family, as well as for the unmarried firemen. Married men had a dormitory and went home to meals in shifts. Eventually, the fire division settled at 40 men working six days and nights a week, with one day off. Married men at the Central Station had apartments in the nearby former Imperial garrison quarters at Fort Townshend, but now the headquarters of the Newfoundland Constabulary. The latter succeeded the Imperial troops in maintaining law and order in the colony on their withdrawal.

The first pros in the Montreal fire brigade were a small, hard fighting crew, one company seen here with their unique devidoir-double, two handreels joined together and pulled by a horse.

(XIII) The Professional Force

Montreal's Guardians

When first hired, the full-time paid guardians, as they were called, simply moved into the barnlike Montreal engine houses, each accompanied by a horse, his wife and his children. They set up housekeeping in the hayloft or whatever corner was available. Conditions improved, notably when room had to be provided for two more men per station in 1863.

The guardians also carried out the duties of watering the streets with watering carts to keep the dust down and lighting the gas streetlamps in the evening. When the alarm bells sounded they had to gallop back to the station to find out where the fire was, just as the drivers from the other stations had to go first to the station where the alarm was sounded to learn the location. So, too, did the volunteers, unless they saw the fire or the reels passing by.

Appointing the guardians meant there was only one lone fireman moving to live in each fire station, but it transformed conditions in the fire department. Now there was a stove and someone to keep it going. Someone was oiling and cleaning the hose and equipment. A horse could be kept on duty and properly cared for. When a citizen came up to give an alarm, the guardian or his wife furiously rang the bell in the tower and the guardian-driver galloped off to the fire with horse and hosereel.

The original guardian for each station was selected from among the company members of that stations's volunteers. The most colourful of these was Jean Naud, founder of the Voltigeurs Volunteer Engine Company in 1848, a huge man of 350 pounds, with a foghorn voice that carried easily above the general din at any fire. He became a legend for his prowess as a firefighter strongman — and for his liquor capacity.

With a turnout of two or three reels and the hook-and-ladder truck, there were three or four men, a bare working crew, but able to attack a fire even if no volunteers turned up. At first, Chief Bertram rode with the guardian at headquarters or went on foot. There was thus a nucleus of strictly serious-minded professionals always quick to arrive on the scene.

The principal remaining cause for delay was now the time taken for the citizen to reach a fire station to sound the alarm, there being no telephones until the 1880s, and few of them then.

Guardian William McRobie said that at night his dog, Carlo, would bark at the sound of anyone running toward the station. He claimed he often entered in his fire report, under First Notification of Fire, the word "Dog." The condition was rectified in 1862 with the installation of the fire alarm telegraph box system.

The city fathers soon saw the protection advantages of the career men in the stations and the next year passed a bylaw setting up the City Fire Police, as the new fire department was to be known. The volunteers were reduced to a small reserve called the City Fire Company. This was composed of three divisions of 12 regulars and six substitutes. They were authorized to wear red, blue or green rubber ponchos according to division. They all received their permanent exemptions from capitation tax, jury and special constable duty when finally disbanded altogether in 1867.

The career force was increased to consist of Chief Bertram, Assistant Chief William Patton, a quartermaster, hose and harness maker, and for each of the eight stations, a guardian, assistant guardian and driver. This force came into being May 1, 1863, and that date is generally taken as the date of the founding of the professional fire department in Montreal. The year also saw the erection of a new headquarters station at the Craig and Chenneville streets site, complete with chief's residence and hose and harness shop.

The Career Department in the Nineteenth Century.

Response to alarms in Montreal in 1863 included two or three stations, each with a horse pulling the unique Montreal four-wheeled double hosereel with 600 feet of hose, plus the ladder cart. In the areas high on the mountainside above Sherbrooke Street, a hand-engine also was dispatched. The men from the remaining stations also responded, but on foot, leaving their guardians behind with their horse and reel in case of other fires or a call for more hose.

The early career firefighters fought their fires without any protective clothing or boots being provided for them and received a salary of $240 per annum. The Montreal department answered 157 alarms in 1863. Of these, 44 were false alarms or chimney fires, and six were listed as "outside the toll gates."

The meagre size of the fire force called for quick, courageous and energetic efforts, a condition that still has its parallels today. In September of 1867, while attacking a fire in the small Spellman Distillery on Longeuil Lane, actually a former church, a wall fell outward, burying fireman William Sharpe of No. 1 and killing him outright. This was the first death in action of a career firefighter and was felt keenly by his fellows. The dangers of the career

Prefire planning by examining buildings under construction was, in the 1860s, one of the progressive orders issued by Alexander Bertram, the first fulltime fire chief in Montreal.

they had embarked upon were reinforced only the next year when another of the small firefighter force, Hugh Scott, along with a helping fire buff, were lost in an explosion at a liquor store. The buff was a popular young reserve army officer named Edward Thompson. One had a fire service funeral and the other a military one.

Fire chief Bertram gradually brought an organized approach to the fire problem in Montreal. Well-travelled and having a mechanical bent, he turned to better equipment to enhance his little force of men. By 1869, from nine stations, he was protecting Canada's largest city, operating nine hosereels and a ladder truck as first line response, with three- and four-man crews. He had two spare ladder carts and was introducing sleigh runners to the apparatus for winter use. No. 5, on St. Catherine Street at the top of the city, had two hand-engines for use in the high outskirts, where there were no hydrants or the pressures were too weak because of the elevation. The last two hand-powered engines in Montreal, they were purchased from Perry Brothers in 1869.

In that year there were 143 fire calls, down from 222 in 1868. The budget had reached $22,236.95. The department was making its riveted leather hose, with 6,250 feet in use. The old volunteer engine houses had now been replaced where unsuitable by new two-storey stations of brick or stone, with apparatus room, stable, bunkroom, and a suite for the guardian and his family. These buildings cost about $2,000 to build. Headquarters on Craig Street, in recent times the site of No. 20 Station, had a handsome coat of arms, a meeting hall, the chief's apartment, hose-drying tower, and hose and harness shop.

In his 1869 annual report Bertram was pushing for more, asking for a steam fire engine for the upper levels and the establishment of a salvage company. He also asked for larger water mains, as the fire potential was increasing with bigger buildings going up in considerable numbers. He got an English Shand Mason steam fire engine for No. 5 Station in 1871, and the three-man salvage company the following year. The chief saw to it that his men had a library, were issued protective leather helmets and conducted periodic drills. A firemen's benevolent association was established with Bertram as president and a group life insurance policy carried.

As the 1870s progressed there was increased communication with other cities, notably major American centres where new developments were being tried and proven. Plain rubber hose replaced riveted leather, but was soon succeeded by the cotton-jacketed, rubber-lined type. The last leather hose was made in the Montreal department's shop in 1876. Leather helmets, still favoured by many firefighters into the 1970s, were first issued in Montreal in 1878. A second Shand Mason was acquired then, though steamers were still rarely used in this city.

Alexander Bertram died in 1875 and was succeeded by his deputy, William Patton. The new chief was apparently an unreasoning disciplinarian, unpopular with the men, and Captain McRobie of the Salvage Corps took early retirement because of what he considered the decline of the department. But it was a time of continued city expansion, the annexation of the eastern suburb of Hochelaga being the first of many that expanded the city boundaries.

The St. Urbain Street Disaster

With its ancient buildings and frequent fires, Montreal has had severe loss of life in its fire service. Firefighting is one of the most dangerous of professions and none have known this better than the firefighters of Montreal.

A great traumatic shock struck the department in the tragedy of the disastrous St. Urbain Street fire of April 29, 1877, when seven members of the department and a former volunteer were killed and others severely injured. Several other dramatic and destructive fires occurred around this time, causing some loss of confidence in the fire department.

It was five in the morning on Sunday, April 29, when the alarm bells struck for a fire at the Oil Cabinet and Novelty Works building on St. Urbain Street at Craig. Woodworking and furniture finishing made for a hazardous occupancy. The building was brick, but with unusually large window areas for the day. This must have made for a weaker structure and increased the draft and radiant heat facing adjoining buildings. There had been a fire there before, and former fire chief Bertram had warned his men that any fire would have to be stopped quickly or lives would be lost if they did not stand well clear.

The Montreal firefighters, small in number, fought stubbornly to contain this new blaze, but were soon forced to retreat hastily into the street and alleys. Here they worked their hoselines from dangerous positions to protect the nearby houses that kept catching fire around their wood trim. William McRobie criticized Chief Patton — for whom he had no respect — for unduly risking his men. The pioneer Skinner aerial ladder was put up but was damaged, and three men on it directing a hosestream burned severely as a sheet of flame burst from windows below them.

The structure weakened as it burned and parts of the five-storey walls began to fall, but the branchmen stood their ground. One crew worked desperately in a side alley to halt the spread of fire into some sheds when part of the north wall crashed down upon them, burying them under bricks and timber. Two had their heads free and could call for help.

Without a moment's hesitation, firemen working on St. Urbain Street dashed into the alley to the aid of their brothers, knowing full well their danger as the remainder of the wall teetered above them. While they dug desperately at the hot bricks with bare hands, the second crash came, burying the rescuers. According to Captain McRobie, they simply flung themselves on top of the trapped men saving the lives of those still living.

A second wave of rescuers came in. They found eight dead, including four firemen, three assisting citizens and the waterworks foreman, a former volunteer. Among the badly injured were two firemen who soon died, two others who never worked again, and the chief. One of five firemen who sacrificed themselves for their buddies was William Ferguson, a veteran fireman assigned to hose repairs because of age, but who had not hung back. To "Old Willie" and his mates, esprit de corps was their way of life.

By 1883 telephones were coming into use and were installed in the fire stations, another great leap forward in communications. A great boon to the poor, long-suffering fire horses was also introduced. This was the drop or swinging harness that relieved the animals of wearing full harness constantly. The first primitive aerial, or mechanically raised ladder, the Skinner ladder of the early seventies was succeeded by the first real aerial ladder, the American Hayes type, as the city core grew into one of four-and five-storey, mansard-roofed blocks.

A disaster for the Montreal fire department was the St. Urbain Street fire of 1877, when seven of its members died. Others were injured as seen in this news magazine drawing as a flash of flame caught them on their primitive Skiner aerial ladder.

Scene of disaster for Montreal firemen. Digging for the bodies of those buried by a falling wall at the Novelty Oil Cabinet Works. Extension of the fire was stopped at great cost to the fire department. - PAC-C65829

At the peak of the horsedrawn apparatus era there were powerful forces available. This big Amoskeag steamer, No. 3, with its beautiful team of horses pumped 1300 gallons per minute.

The increased heights dictated, as well, the use of steam fire engines downtown to supply adequate hosestreams on the upper floors, so Montreal joined other cities in the general use of these to supplement hydrant pressures. The first of three Merryweather engines was added in June of 1886.

The Boy from St. Chrysostome

An era of enlightenment dawned with the replacement of Patton on his retirement in 1888. The new chief was a former grocer's clerk from St. Chrysostome village, Zepherin Benoit, who had lived part of his boyhood and was educated in New England. Returning to Montreal, he served first as a policeman and then in the fire department. This experience put him in a good position to take on the job of organizing a police-fire department when applications were called for in the neighbouring suburb of St. Henri. He also had the recommendation of Chief Bertram for this job.

When Patton retired, the 38-year-old Benoit had earned a reputation that gained him the invitation to command the Montreal department. Calamitous fires of the preceding 20 years had shown the need for modern and inventive ideas. These disasters included the St. Urbain Street fire, the St. James Hotel fire, the Ostell-Shearer Lumberyard and St. Patrick's Hall fires, where criticisms had been leveled at the fire department and at city hall for weak building regulations.

Benoit proved to be the right man. He introduced chemical engines, bigger steam fire engines, and endorsed the appointment of two city fire commissioners, new positions to provide for the investigation of fires. Fred Perry was one of these. Benoit became the first Canadian president of the International Association of Fire Engineers and led a Canadian delegation to the great International Firemen's Tournament in London, England, in 1894. In addition to knowing what is needed, a fire chief has to convince city hall, too. Benoit had a great ally in the fire committee chairman, Lieutenant-Colonel A.A. Stevenson, a keen student of fire protection. Some city councilors decried the chief's "shabby tactics" in June 1906,

Horses in fires were a considerable problem to the firemen as they would panic or persist in seeking supposed security in their stalls. Fire in Bancroft Livery Stables, Montreal, 1875.

however, when he asked permission to put his men of Station 20 in tents at night, as they were too nervous to sleep in the old building with its bulging walls and falling bricks. But, they got the point and Benoit got his new station.

Benoit led the department in an era of vigorous expansion as various suburbs were gobbled up. By 1890 there were 15 fire stations housing three aerial ladder trucks, eight steamers, 15 hosereels, two chemical engines, three salvage wagons and several city service ladder trucks. These were manned by a captain and three branchmen for hose crews, an engineer and driver on steamers, a foreman and four laddermen on ladder trucks, three men on chemical engines and four on salvage wagons. By the turn of the century there were 100 horses in the department and double-box hosewagons were replacing the traditional hosereels.

The first recommendations for a fire prevention specialist in 1890, and for a training school in 1903, came from Benoit, but the changes did not come in his time. The Board of Trade conflagration in 1901 brought pressure from the fire insurance underwriters for still more and better apparatus. It was about this time that the fire insurance companies began joint inspections of municipal fire defenses through their provincial rating associations.

Zepherin Benoit retired in 1909, having moved the Montreal Fire Department into the high gear of progress and expansion ably continued through World War I by his successor. This was Joseph Tremblay, his successor in St. Henri, too, who had become a district chief in Montreal when St. Henri and adjoining St. Cunegonde were annexed in 1905.

Toronto's Rapid Expansion

The Toronto city fathers had considered steam fire engines as early as 1858, the fifties having been a period of increasingly serious fires in a growing and congested commercial area. Two of the unique Silsby rotary pump engines were brought across from upper New York State and put in service in 1861. No horses were arranged for, so they were often drawn by hand like the lighter hand-engines. Although Toronto was fairly flat, the mud was often rather cloying. Usually the firemen searched the streets for a passing horse to commandeer. There was jealousy on the part of the remaining hand-engine crews and some cases of hindrance of the steamers occurred.

With the great power of the two steam fire engines established, another reorganization of the fire department was inevitable and this came in 1862. This time it was a drastic cut eliminating the large volunteer companies and reducing the department back to two stations, No. 1 on Court Street and No. 2 at Bay and Temperance.

A small on-call paid force was set up,consisting of Chief Engineer Ashfield, Assistant Chief Engineer Richard Ardagh; an engineer, driver and stoker for each of the two steamers; two hosecart drivers; a hook-and-ladder cart driver; 17 branchmen to handle the hoselines; a bugler (rather unusual); and others to make up a total of 40.

These firemen apparently all had other jobs but were paid to respond to all fires thus falling into the on-call category. There was no one on duty in the two firehalls. Three hand-engines were retained for a short time into 1862, and the others sold off to various small towns. That of the Phoenix Company became nearby Oakville's first engine.

To supplement the rather limited piped waterworks, 28 street tanks or cisterns were built to supply the steamers. Three horse-drawn hosecarts were put into service. These were two-wheelers carrying a reel of hose, typical of the era. Seven horses were taken on to provide for prompt responses with the heavier equipment now in use. Toronto had 55 fires and 25 false alarms reported in 1862, with a total loss of $47,000. Incendiarism was suspected in 15 of the fires. In all, it was a pretty good year. The department now had 2,400 feet of hose and a fuel wagon to replenish the steamers at fires.

Some thought to the heroic efforts and dangers of the firefighting trade resulted in recognition by the city in 1867, when provision was made for "a medal or pecuniary assistance for distinguished service," as well as awards to widows and orphans.

The horsedrawn hosereel ran out the company's hoselines from the steamer or a hydrant. This is a Windsor crew.

Toronto was now growing in size and prosperity as the political and commercial capital of Upper Canada, about to become Ontario in the new Confederation. The slow, elementary means of spreading the alarm of fire by hot-footing it to the nearest firehall was replaced finally in 1870 with the modern electrical telegraph system, already in use in Saint John, Montreal and a number of American cities.

In 1871 the Berkeley and Bay Street firehalls were enlarged and new ones built on Yonge and Portland streets. These were completed to replace the sadly inadequate old volunteer engine houses and to accommodate four steamers, four hosereels, a hook-and-ladder truck and, in 1875, a salvage wagon. The fourth engine was another Silsby and was named James B. Boustead in honour of the chairman of the Fire, Water and Gas Committee, a great booster of the fire department.

In this year, Chief Ashfield had a 50-man force, 11 of whom were full-time staff, and included the engineers and stokers, on continuous duty in the stations. The others were on call. The bulk of the membership worked at other jobs scattered about the city and slept at home resulting in some having difficulty hearing the alarm bells. They were also evidently not under compulsion to attend fires unless it suited them.

As elsewhere, the advent of steamers caused a drastic cut or elimination of the hand fire engines in Toronto. Only two stations remained, each equipped with a Silsby steamer.

Turning Pro

For several years Ashfield had been eyeing the highly reputed Montreal Fire Department, now a totally career force. The Fire Committee visited Montreal and was impressed with the efficiency of the fully manned stations. In another three years, the committee reported that the existing system could not provide proper firefighting service as should be expected in an important city. The city fathers agreed and in 1875 ordered the setting up of a revamped department based entirely on a force on duty at all times in the four fire stations.

The new Toronto professional fire department personnel consisted of the chief, assistant chief, alarm electrician and his assistant, an engineer and a stoker for each of the four engines, seven foremen and 26 firefighters. In addition, there were nine teamsters and 13 horses retained under contract. Four hose-and-engine, one hose and one ladder companies were now being manned. The new career firefighters lived in the stations and had "a few hours off once in a while."

The next year, Chief Ashfield, who had joined as an original member of the Phoenix Volunteer hand-engine company in 1839, was unable to carry on the active work of fire scene command, presumably for reasons of health. While retaining his title and administrative duties, he turned over the job of operational chief to his assistant, Richard Ardagh, who had joined as a volunteer at the age of 15 and now assumed the title "Chief of Brigade."

The number of fire stations, or halls as they are often called, now grew rapidly with the expanding city, annexations making up a good part of the growth. To the original manned stations on Bay, Portland, Yonge and Berkeley streets, a hose station on Queen and the "ladder house" on Court Street were added. In 1878 stations seven, eight and nine on Wilton, Dundas and College were also added.

The fire station of Yorkville was annexed in 1885 and two more built in the same year, the department now numbered 74 men and the original 20 street alarm boxes had become 154. This last statistic reflected the greatly increased area under protection. The chief now received an annual salary of $1,000 and the firefighters $480. Ontario was becoming great industrially in the post-confederation period, the strategic hub of a country filling in from coast to coast. This worked to the disadvantage of the Maritimes' cities, which suffered a decline in growth. The development of the respective fire departments can be seen to have followed this pattern.

Chief Ardagh, who took over fully from Ashfield in 1888, repeatedly objected to the city's reduction of the number of steamers in service, but in vain. They were, in fact, all retired in favour of the new chemical engines and direct hydrant hoselines. Ardagh was given the city's first aerial ladder truck, an 85-foot-high, hand-cranked ladder by the Babcock Company, but it is doubtful the hydrant pressures could have supplied it an effective nozzle stream at any great height. It ran out of the downtown Lombard St. firehall.

Normal hydrants pressures were 80 pounds or less in the business district and half that in the higher levels of town, barely enough to handle a serious house fire, let alone one in a large mercantile building. The arrogant overruling of the experienced officer they paid to be their expert advisor turned out to be a grievous error on the part of the Toronto city administration. Their folly was to be clearly demonstrated in a few years with the disastrous *Globe* newspaper fire of 1895. They even had a warning in the Esplanade Street conflagration, August 3, 1885, which razed the grape-sugar factory, elevators, shipping and other buildings along the waterfront from Princess Street to Church Street.

In the meantime, some progress was being made toward improvements in the welfare of the now well-established career firefighter. In 1887 a new bylaw provided that each man should receive one half-day and the following night off duty each week, a great breakthrough. This period ran from one p.m. to eight a.m. the next morning. It was similar to the procedures gradually being introduced elsewhere. It was also the general rule for the men to go home to meals, in shifts, during which periods they still had to respond if an alarm sounded.

At this time formal entrance requirements were introduced, calling for a medical examination, a maximum age of 26 years, and an ability to read and write. Three years later a superannuation and benefit fund was set up, the city funding it with $10,000 and a payroll deduction of two percent.

The Globe, Osgoodby and Simpson's Fires

Several serious fires in the increasingly tall buildings in 1894 presaged more serious difficulties for a fire department with such weak firepower. Chief Ardagh asked for steam pumping engines every year, but got only hose and chemical carts. The dramatic cases that finally sold Ardagh's arguments to council came at the beginning of 1895, but the frustrated chief did not live to have the satisfaction of seeing it happen — losing his life as a result of one of the fires that convinced the politicians.

It was at 3:30 a.m., January 6, 1895, that alarm box 21 struck on the bells, signaling a fire at the corner of King and Yonge streets. When the first fire companies galloped up, fire had spread from cellar to roof of the Toronto *Globe* newspaper building through the vertical shafts. In a raging blizzard, the puny streams of the chemical engines and low-pressure hydrant hosestreams could make no impression on the inferno. They could not reach the upper floors of the five-storey structure.

The building's interior quickly burned out, a wall fell into Melinda Street, killing Fireman Robert Bowery and wrecking the aerial ladder truck. Leading his men in an interior attack, Chief Ardagh and two men were forced to leap from a window as the floors collapsed. All were injured, and Chief Ardagh died three weeks later.

Four days after the *Globe* debacle, another downtown edifice, the five-storey Osgoodby Building, and several warehouses burned out. Three occupants were saved sliding down wires or jumping into life nets. There was now no aerial ladder available, though the wrecked Babcock truck would eventually be rebuilt. The lack of power to control well-advanced fires was clearly shown. Then, on March 3, Simpson's department store burned in a spectacular inferno whose flying embers reached the Don River, a mile away. Several neighbouring buildings were lost, as well as the steeple of the Presbyterian church. The block-long Confederation Life Building across Yonge Street, was saved by its own interior hoselines, surviving to become a handsome heritage edifice in the 1980s.All three fires spread rapidly because of open vertical shafts and were stopped more by snow-covered roofs and solid walls than by firefighting.

These three fires were lesson enough for the Toronto city council. A big new 1,000-gallon-per-minute steam fire engine was received as early as April, another later in the year, and the old James Boustead engine was overhauled and returned to service. Two new hosewagons and a 65-foot Champion watertower, for delivering large streams of water into the upper floors of large buildings, were also purchased. Ardagh's successor, Thomas Graham, urged the council along in this rehabilitation, though he remained in office only two years.

The growing strength of the Toronto Fire Department was a great help in the battle against fires in the ever-larger structures stuffed with the combustible goods of a prospering country. One serious fire protection weakness, however, was growing steadily into a greater danger, since it was not keeping up with either the fire department or the fire potential. This was the water supply system. Again, the city would have to learn the hard way, from one of Canada's most costly conflagrations, that of 1904 (described later). The Toronto Fire Department had become a professional or career force, in keeping with the growth of the city it protected, but provision of all the tools it needed to do its job still tended to lag behind.

For big fires, even the strongest domestic waterworks systems became insufficient. The steam fire engine developed the high pressure hosestreams needed for high and deep mercantile buildings.

Inability to supply powerful water tower streams like this was a factor in three large fires in downtown Toronto in 1895.

Ottawa Reforms Its Fire Department

As lusty Bytown faded into history and the growing political and lumbering capital of Ottawa took over suburbs like New Edinburgh in 1877 and Rochesterville in 1890, pressures came for improvements in the fire department. In 1874 Chief William Young and 18 career firefighters were installed in the stations to commence a phase-out of the volunteers.

Steam fire engines, of course, made this possible, along with the finally completed waterworks and hydrant system in 1872.

However, the initial key to a successful career fire service was copying the example of Montreal. After years of dithering, a strong waterworks was built, in which the water was pumped with watermotors run by the river drop at Chaudiere. The delays in construction prevented the present Parliament Hill from becoming the site of a large water reservoir. Efficiency, for modest-sized fires at least, was secured by fast action with direct hydrant streams in answer to quick notification via an electrical fire alarm telegraph box system by Gamewell.

Ottawa had not had a good reputation with the fire insurance companies and there had been complaints over the high rates. The city and its neighbour, Hull, across the river, had grown rapidly without much effort to control highly combustible construction.

Trying to come to grips with the fire problem, the city brought in Peter Prevost, a member of the Montreal Fire Department, in 1897, to emulate the bigger city's highly reputed techniques. Prevost had a free hand in transforming the fire department. Several older officers were retired and young men promoted, with old apparatus replaced as well. This was all triggered, as usual, by a series of serious fires. Chief Prevost led the department until his death in 1909.

In Chief Prevost's new department, the spectacular showpiece and pride of the department was the Langevin aerial ladder and its well-drilled crew. A Montreal invention, this was one of the more successful early attempts to build a truck-mounted mechanical ladder. All these improvements would prove useless in the face of huge conflagrations at the turn of the century.

Hull Goes the Quebec Way

Though a smaller city directly in the shadow of the capital of Canada, Hull evolved its fire service in the mode of Quebec towns, generally. In 1875 it became a city, and four years later a police department was formed. Typically, the organization of the volunteer firemen was placed under the charge of Chief of Police Ludger Genest.

This was not an original French Canadian concept, but as clearly seen in opposite ends of the country, Victoria and St. John's it was the British idea of direct municipal control of a municipal service through an established, full-time uniformed official. Chief Genest formed and became chief of what was named the "Jacques Cartier Volunteer Fire Brigade" in 1885. It had a membership of 30.

There followed shortly the construction of a waterworks system to replace the carters and their water barrels. A horse was purchased and a force of four police constables hired. These men doubled, of course, as firefighters. First headquarters was on Duke Street. The City of Hull was an uncontrolled wooden mass, mostly homes of workers from the big wood-processing mills at the Chaudiere, and it lived on luck.

It was not until 1898 that a separate fire chief was appointed, separating, for a time, the police and fire services. Eight full-time firefighters were retained. The reason for this separation is not known, lasting only four years, as it did. It was in the time of this tiny separate department that the devastating Ottawa-Hull conflagration of April 1900 wiped out much of Hull, including the fire station, and a sizeable part of Ottawa. Two years later the police and fire departments were again amalgamated, with Chief of Police Genest as administrator and a fire chief under him. A steam fire engine was purchased and more firemen engaged. This structure was to remain for a half century more before another separation of the two forces.

Again, this was typical in the province of Quebec, the firefighting operation coming under the chief of police, whether volunteer, policemen performing double roles, or a variety of mixtures. In recent times, as cities grow larger, partial or total separation of the two functions has been occurring.

Calgary, seen on the day of its first serious fire, which occurred on Atlantic Avenue, January 8, 1885. Buildings are typical of the new towns in the west.
– PAC-C56992

4

OPENING OF THE WEST (1860-1890)
(XIV) Prairie Cities

By the last quarter of the nineteenth century, Eastern Canada was a well-established society, cities and towns tied with roads, railways and canals. An agricultural economy it was, but fairly self-sufficient and comfortably civilized compared to the wilderness of the preceding century. To the west were the Great Lakes and then the formidable bush and rock barrier of the Canadian Shield, on the far side of which lay the empty frontier lands of the Prairies and the western mountain ranges. Except for a few coastal communities on the Pacific, all this was barely explored, let alone settled. Civilization was represented across the vastnesses of northwestern Ontario, the plains and the mountain country by a few forts and sod-breaking settlements, often hundreds of miles apart.

In 1872 the engineer Sandford Fleming made his three-month trek from Toronto overland to Victoria by ship, canoe, wagon and horseback, assessing the best route upon which to build the Canadian Pacific Railway. Traversing the wilderness with guides, his small party passed through military outposts, religious missions, Indian encampments, Hudson's Bay Company posts and a few hardy farm colonies, but there were as yet no towns worthy of the name between Prince Arthur's Landing, at the head of Lake Superior, and the Pacific coast. The Landing was a hundred shanties, houses and stores. A farm colony of 15,000 was scattered about Fort Garry. There were then the Fort of Edmonton and the goldminers' bases in the Cariboo, on the other side of a vast prairie.

Sandford Fleming's diarist on his monumental expedition, Reverend George Grant, commented on the frequency of Prairie grass fires and bemoaned the profligate way in which travellers and nomadic venturers on the plains in the 1870s caused them. These fires in turn burned off the modest amount of wood that grew, thus reducing the moisture content of the soil and noticeably depleting the water supplies. Previously wooded areas were now bare, he observed. Grant hoped that settlement of the country would soon inhibit the carelessness with campfires that was the root of the problem.

The first fire engines on the Canadian Prairies would in all probability have been imported by the Hudson's Bay Company to protect its posts, and the first may well have been the one at Fort Garry, at Winnipeg. An old engine, last seen disappearing in the vegetation on the bank of the Red River, is believed to be this machine. A photograph of the remains indicates that it was a Hunneman engine from New England, probably brought in by the river route from the United States about mid-century.

Like the military garrison engines used by British troops in Eastern Canada, the Hudson's Bay Company engine was sometimes run outside the company's properties to fight town fires. However, there seems little indication the community was thus inspired to acquire its own fire engine. Development of Winnipeg to such an organizational stage did not take place until it became a city in 1874 and the steam fire engine had displaced the muscle-powered hand-engine.

The Prairie towns opened up with a rush when the C.P.R. pushed across the flat land, replacing the wagon trails. European immigrants poured in, hopeful and naive. Winnipeg was already a trading point, with river traffic from an American railhead south of the border. Settlers were developing the rich surrounding farmlands and the town was finally chosen as a centre of construction operations for the great C.P.R. project.

Winnipeg formed a volunteer fire brigade through a bylaw passed in September 1874, soon after the city was incorporated. Now in the steam era, a Silsby steam fire engine, four hand hosereels, a hook-and-ladder cart, and 2,000 feet of hose were ordered. A number of the customary pre-waterworks street cisterns were built as suction sources for the engine.

Firemen and visitors gather at Winnipeg's temporary first fire station after the original burned. The short fireman with the beard is "Fighting Billy Code", a firefighting cat-with-nine-lives.

The Winnipeg Volunteers

G.U. White was elected first chief of the Winnipeg volunteers. As in most towns, among the first volunteers were some of the most prominent citizens, and the names of the later great families were on the roster — Ashdown, McKechnie, Fonseca, McMillan and others. There were apparently a hose-and-engine company and a hook-and-ladder company at the start. The six water tanks were located along Main Street, the heart of the city, and these were filled by the steam fire engine from the river. A site on Post Office Street, later renamed Lombard Avenue, was chosen for the first fire station.

The town of such promise seemed to have its trials with operating a fire department in the earliest days, however. The first steam fire engine arrived, like the first railway locomotive, by barge on the Red River. A goodly crowd gathered for the first test and demonstration of the powerful machine on November 28, 1874, and it went into service, proving a potent weapon against the dreaded scourge of fire. However the original hook-and-ladder truck had been lost when the train carrying it plunged through a bridge in Minnesota. Delivery of the other equipment, arriving by riverboat, was held up in a squabble over freight charges.

Then, on Christmas morning 1875, the firehall caught fire, burning so quickly the firemen barely escaped, leaving their steamer to the not-so-tender mercies of the vengeful flames. The town was left without fire protection. An emergency meeting of council immediately telegraphed the Silsby Company to rush a replacement engine and 2,000 feet of hose The next year, disagreements between the firemen and the council led the firemen to resign. The department was reorganized the next month, but a year later, there was yet another reorganization and a new chief appointed in an attempt to secure harmony.

Under the new chief, Dan McMillan, later lieutenant governor of the province and knighted, a new station was erected on Market Square. The station was opened with an oyster supper. The new, bigger steamer was christened Assiniboine and, on being rebuilt by Fire Lieutenant John McKechnie, founder of the Vulcan Iron Works, the old burned engine was appropriately named Phoenix. New captain of the Hose and Engine Company was a tough little Irishman named Billy Code, destined to go down in local history as one of the colourful, nine-lived firefighters that seem to occur one to a department.

The Grace Methodist Church bell was used as the general fire alarm. Volunteers were paid a dollar for each fire they fought and conversely, fined a dollar for nonattendance. The new fire department had its first serious test that May, the burning of the Dominion House and its stables. Hotels of that era naturally provided accommodation for the guests' horses, too. Starting in the wee small hours, the flames destroyed the buildings, but the department made a brave stand and preserved the adjoining Garry House. It was now a 35-man force.

A new chief, Alex Brown, took the reins in late 1877, carrying on until the volunteers disbanded in favour of a career force in 1882. In the interim a severe winter fire on Main Street, when the street tanks were frozen, and another towards Spring demonstrated the need for a stronger department. Two new engines, another Silsby and the unique two-cylinder Ronald, were bought. The Ronald was restored to running order for show purposes by the department's personnel in 1982, one hundred years later.

Times of McRobie and Code

The new full-time department was established and placed under the direction of an experienced officer, none other than William Orme McRobie, former captain of the Montreal Salvage Corps, at a salary of $1,800 and living quarters in the central fire station. He was progressive and brought experience and professionalism to the organization. Like Billy Code, he compensated for his short stature with toughness, energy and courage.

A notable fire in the first year of the career fire department was the burning on October 10 of the offices of the Canadian Pacific Railway Company and the adjoining Bank of Montreal. Fire was discovered in the evening, issuing from the ceiling below the offices of William Van Horne, who was building this great transcontinental railway (as a result of this fire he lost all of his papers). Chief McRobie and his men emptied the 28,000-gallon tank at Portage Avenue to control the fire.

At this outbreak, primitive spray nozzles were used for the first time by the department, as were relief valves to prevent excessive pressures on the hoselines. These new devices were introduced by McRobie, who also brought in a system of gold, silver and bronze medals for heroism in the fire department, and who liked to test the response of his men by pulling false alarms on the alarm boxes.

McRobie, Code and others were recognized for courage in fighting the Ashdown Hardware warehouse blaze, containing a considerable stock of gunpowder. Several firemen, including McRobie and his son, as well as some bystanders, were injured in the ensuing explosions. While some kegs of powder were going off, they carried others to safety. The Firemen's Benevolent Fund benefited as a result of this illustration of the job's dangers.

The city was expanding as the "Gateway to the West." The new Central Fire Station

William McRobie in the uniform of Captain of the Salvage Corps in Montreal before he moved on to reorganize the Winnipeg Fire Department following an active career in the older city's fire department.

A unique uniform was chosen by the firemen of Prince Albert, Saskatchewan. They pose proudly before their station in 1907.

was opened in January 1883 on William Avenue, and shortly after, No. 3 was opened. The new department had 36 full-time firefighters, 17 horses, four steamers, three chemical engines, three hosewagons, a ladder truck and 8,700 feet of hose. Temporary headquarters at Smith and York streets, known as South Fire Hall, now became No. 2. Like many busy fire stations, it had for many years a high degree of esprit de corps.

Full-time was really the word for the firefighters of that period. They lived in their fire stations except to go out in rotation for meals. Single men had one afternoon and evening off per week, while the married men had also a full night off each week. But the department had grown and improved greatly in its first decade.

Chief McRobie retired from the scene in 1889, having brought organization and new ideas in a way the procession of untrained volunteer chiefs had not. He had the advantage of his 27 years experience as both a volunteer and a career firefighter in the older, established fire department of Montreal. He had his medals to attest to his bravery, too, but must also have had his enemies. He was once accused of drunkenness, but was eventually exonerated by city council.

As successor, the colourful "Fighting Billy" Code, who had become deputy chief, was made acting chief of the department. Apparently the press urged council not to appoint him permanently until he demonstrated his ability to handle the top job. It only took a major fire involving several buildings and featuring a daring rescue to settle the question in Code's favour. Noted for his derring-do, perhaps the little Irishman did not exhibit the more polished characteristics normally expected of an administrator.

William G. Code had emigrated from Ireland to Michigan in 1872 eventually making his way by boat, horseback and on foot to Winnipeg to work as a bookbinder. He arrived in time to join the newly-forming volunteer fire department, having had an interest from his childhood in the old country, where he had lived across the street from a fire station.

When the budding career fire department was formed, Code was quick to apply, and he became a foreman. Throughout his career, Billy Code was one of those firefighters who showed no fear and was always in the hottest spot. He carted kegs of gunpowder from the blazing Ashdown warehouse, considered it memorably cold at -53 degrees Fahrenheit during the Manitoba Hotel fire in 1899, and was frozen to the street at another fire — his men had to break him loose and carry him into a nearby hotel to thaw out. He was crushed by falling timbers, trampled by horses and thrown about by explosions. His small beard was worn to hide scars of past fire injuries.

Fighting Billy drew additional public attention by dashing off with detachments of men and equipment to answer pleas for help in other towns. He took his flying forces by special train to fight forest fires at Rainy River and even across the Minnesota border when assistance was asked.

Confirmed as chief, Code served as the second career chief in Winnipeg until 1895, when he stepped down again to the rank of station captain. Old correspondence includes letters between Code and Captain E.H. Rodgers in which their titles were switched, the latter notifying Code he was to be considered second in command of the department. No reference to the reason for the switch was included.

After his six years as chief, Code went on to work another 26 years as assistant chief, serving as a firefighter for the Winnipeg Fire Department 44 years in all. He lived on until the age of 92, being buried with full honours in 1940.

Chief McRobie had served in Montreal, where a street fire alarm telegraph system had been installed in the early 1860s, vastly improving the speed and accuracy of transmitting alarms from the public. He was probably pleased to have such a system completed in Winnipeg in the year he took charge. Twenty-four street alarm boxes, accessible to the public, went into service March 23, 1882, with six gongs sounding the signals in the fire stations and other key locations.

A Portable Town

Further west, a place known as "Pile of Bones," a muddy spot with no wood, no water, and no apparent reason to become a town, blossomed into Regina, the capital of the Northwest Territories in 1883. To the north, possibly a more attractive site, was Saskatoon, but it stayed a struggling village because it did not have the first great railway come through town.

Since the site of Regina seemed to have little reason for becoming a city other than the expectation of the coming of the C.P.R., the town that was built in anticipation of this event was somewhat portable. There was insufficient confidence as to exactly where the railway would lie; many buildings put up in anticipation of the boom were temporary, portable or even of canvas construction. The town eventually became established as an appendage of the rail line, but must have remained a disorganized settlement with little control for some time.

In 1886 lack of street lighting, refuse lying about and seemingly shabby treatment of the volunteer fire brigade were the order of the day. There was no town constable and the couple dozen firemen were discouraged by lack of public support and equipment. Proposals to spend money on improvements in fire protection in the latter decades of the nineteenth century were rejected by the citizenry, so that fire protection in the new capital of the Northwest Territories languished. Major fires in 1890 and 1892 were the result. It took a long time before the Queen City took fire protection seriously.

Fight Over the Firehall

Hardly worthy of a name, the tiny outpost on the east bank of the Elbow River was simply called The Elbow. Then it was named Fort Brisebois briefly before becoming Calgary. In 1881 it was composed largely of two stores and a Royal Canadian Mounted Police barracks and had a population of 75. It began to grow with the promise of a transcontinental railway. Casual firefighting with a few buckets gave way to an organized bucket brigade and hook-and-ladder company, with that Prairie substitute for the hand tub, the soda-acid chemical cart. The number of volunteer members was small, only about ten. The first leader of the force to last more than a week was Captain Steve Jarrett, elected by the members. This was 1885, the population about 1,000.

The first recorded fire in Calgary occurred on January 8 of that year in a house on Atlantic Avenue. Attempts to extinguish this fire with buckets and two water carts, supplemented with snowballs, failed and attention was turned to saving as much of the furniture as possible. Fortunately the mayor and council turned out and saw the need for more serious fire protection measures, hence the organization under Captain Jarrett. This would still be insufficient, however, as was made clear by a large fire the next year.

The centre of the business district of this western frontier-style town, some 14 log and wood-frame buildings, was burned out on October 15, 1886, the year the first train from Montreal to Vancouver passed through. It started in a shack near 9th Avenue and Centre Street, spreading without effective opposition from the volunteers. A new chemical engine and hosereel had arrived but were under lock and key because the town could not pay for them yet. After some deliberation the volunteers broke in and pressed the new apparatus into service.

Following this fire a reorganization of the firefighters took place. Captain Jarrett of the volunteer brigade became fire chief of the Calgary Volunteer Fire Department. The council placed an order with J.D. Ronald for two hand-drawn hosereels, 2,000 feet of hose, and a steam fire engine. Wells had been bored at principal intersections and large wooden pumps provided for filling buckets, but now cisterns were built to meet the larger demands of the steamer.

Construction of the first firehall, on 7th Avenue between Centre Street and 1st Street SE, was approved and opened the next May with a concert and dance. Citizens donated funds to help furnish the building. The new firehall was a commanding structure, towering over the small buildings around, and its belltower was a landmark. The building was so impressive that it created a serious problem. The town council decided to hold their meetings and do their business there.

The firemen took exception to the city fathers occupying what should have been their meeting and club room. The ensuing altercation reached the point where the men and their chief all resigned, selling off their furniture and donating the proceeds to the hospital. Undeterred, council appointed a new fire chief and firemen. They may have had some difficulty, as the new force included the mayor and council members. It proved unsatisfactory, popular demand causing reinstatement of the original firemen, who regained possession of their hall.

Again, in 1895, the firemen threatened to resign until improvements in equipment and provision of horses were agreed to. Previous to this, the heavier apparatus now in service was pulled by horses of any public carters who showed up at the firehall on the sounding of the fire alarm.

As in the eastern cities in their early days, the firemen were very prominent in social events and entertainment in the community. They were involved in the parades, balls and athletic contests. The Calgary Fire Department Band was organized in 1890, using instruments, music and uniforms purchased from the Oddfellows Lodge. It won the Northwest Band Competition two years later.

Another special Calgary event was a race in which competing firemen leaped from beds set beside the racetrack at Victoria Park, dressed and dashed around the track with horses and apparatus. This simulated turnout on an alarm could have inspired the modern chuckwagon race! An event of long-term significance was the succession of the colourful James "Cappy" Smart to the position of fire chief in 1895. He was to lead the department for 35 years, molding it to his own style.

A Late Starter

Though established very early as the outpost Fort Edmonton, the eventual Town of Edmonton joined the western boom later than those towns located along the C.P.R. transcontinental route. Its volunteer fire department was formed in 1892 under R.G. Davidson, a former Sherbrooke, Quebec, fireman. Previously, every citizen turned out with a bucket on the alarm of fire.

It was in June of that year that a group of citizens met at the office of Mr. Daly, who was in the drug business, to elect officers to "manage and conduct the chemical engine, hook and ladder and general apparatus" just purchased by the town council. Thus the Edmonton Fire Department started off, in Prairie fashion, with a double 40-gallon soda-acid chemical engine and a hook-and-ladder cart stored in a shed.

The attributes of the chemical engine were speed and a self-contained water supply, but it was capable of extinguishing only a very small fire or of protecting adjacent structures from embers. So the next year a steam fire engine and two hand-drawn reels were purchased and three cisterns built for the steamer to draft from. This allowed the firemen to bring to bear substantial streams of water from two-and-a-half-inch-diameter hoselines, a serious firefighting capability. There were eventually 11 cisterns or street tanks of 12,000 gallons capacity each. They were filled by the steamer and hoselines from the river below the town.

No. 1 Fire Hall was built at the same time on Fraser Avenue at Rice Street. It provided a dormitory to accommodate 12 firemen, with "free light, heat and water." These retained men had other work, but were paid 60 cents an hour for fighting fires. By 1898 there were 40 volunteer or call firemen and a driver for the steam fire engine's horses under a part-paid fire chief.

The Prairie cities were destined to be born later than those of the East, but in the population explosions of the later immigration waves, to grow in great spurts. Lacking the long traditions, they were quick to pick up on new ideas in fire protection, such as the introduction of the automobile fire engine.

Cold is a severe handicap to firefighting on the Prairies, where the Canadian winters blow fierce. Firefighters and apparatus often become sheathed in ice.

Cold Country for Firefighting

Fighting fires in the severe cold of Prairie winters experienced in Winnipeg, Bienfait and points beyond, has called for great stamina on the part of the firefighter. The human energy consumed working in such low temperatures calls for frequent relief of crews in action, quite aside from the cold. High winds help the fires along and even heavy, high-velocity streams of water are turned to spray which freezes before it can reach the fire.

Cases like Billy Code freezing immobile on the street at the Sterling Building fire and the Manitoba Hotel fire (fought at -54 degrees Fahrenheit) are no exaggeration. Hoselines froze solid in a moment if the water stopped flowing, and aerial ladders encrusted in frozen spray snapped under the weight of the ice. The steamers were useful following winter fires for thawing out ice-covered apparatus and hose so it could be transported back to the station. Horses were taken to the nearest fire station in such weather, though each had a blanket provided.

Other parts of the country had and have the hazards of sub-zero temperatures, but none so severe as the Prairies and the northern territories. Quebec towns have their deep snow, but the Prairies the fast freezing wind. As any experienced fireman can confirm, it was a good idea to get a coating of ice over your turnout coat right at the start, since it formed a seal against that penetrating wind.

(XV) Beyond the Mountains

The First Towns

On the West Coast, where seaborne commerce with the well-established American ports to the south was possible, there were several growing towns around mid-century. Victoria (population 5,000) and New Westminster (population 1,000) were rough yet but had recognizable ambitions to become important cities. These populations had been greatly inflated in the early 1850s by the gold rush crowd lured to the British Columbia goldfields. When the gold rush petered out, the populations and ambitions suffered a setback.

Victoria, with adjoining Esquimault, comprised the capital of the colony of Vancouver Island. At the peak of the gold rush the population was inflated to 15,000 people, mostly transients living in tents and makeshift shacks. The town was prosperous but expensive. Nevertheless, Victoria had achieved some solid progress as an outpost of basic comforts, commerce and government, having easy connections by sea to San Francisco and the world. The San Francisco connection was to have a distinct influence on fire service in Victoria.

Volunteers of Victoria

As the first important town on the west coast of Canada, and under the stimulation of the gold rush boom, Victoria made the first moves in fire protection for the public. Initiatives came first from the citizenry, then the colonial government and finally the municipal authority, with clear differences of opinion as to the basis of control. It started in the early days of the gold rush period, in 1858, when an influx of hopefuls poured into the town from down the American coast and flimsy wooden shacks and tents sprouted up on all sides. The established property owners, becoming concerned over the apparent fire hazards, requested that the governor of Vancouver Island, Sir James Douglas, take action to protect the town from fires.

The story of the volunteer fire department era in Victoria is very well recorded, thanks to the fine research by F.W.Laing and W. Kaye Lamb in their paper "Fire Companies of Old Victoria." The development of the municipal fire service in the remoteness of the British Columbia coast, beginning at Victoria, had unique aspects of particular interest. These included the strong influence of the American gold rush throngs from California, with antecedents on the eastern seaboard. Countering this was the resistance to their ideas by the British governor, who saw a fire service under municipal control, not independent, as in the American tradition.

Governor Douglas responded to the citizens' demand quickly enough, arranging through agents for the purchase of two hand fire engines from San Francisco. Just as eastern towns started off with secondhand engines from older, bigger places, San Francisco engines found their way into British Columbia. One of the first Victoria engines was one of these, its history being traced by Laing and Lamb. Built by John Rogers of Baltimore, it was one of three brought out to San Francisco in 1850 upon the organization of the city's Monumental Engine Company, which included many ex-Baltimoreans in its ranks.

The second engine was a new second-class Hunneman from Boston, painted in carmine with gold striping. It had a hosecart, and 1,000 feet of hose was bought to go with it.

Governor Douglas was faced with a total bill for engines and incidentals of just over $5,000, but this was for some reason paid by the Hudson's Bay Company and ownership only transferred some time later to the colony of Vancouver Island.

Perhaps Douglas thought he had done what was necessary at this point when he put the apparatus in the care of Police Magistrate A.F. Pemberton, but nothing had yet been done about organizing a firefighting force. The shopkeepers and property owners did not let it ride, however, sending a petition immediately to Douglas asking for the formation of a volunteer fire department. Then, without waiting for a reply, they held a meeting in the warehouse of the Hudson's Bay Company, at which the formation of volunteer companies was endorsed and committees set up to organize them, receiving applications for membership.

Organization proceeded apace for a short time, but then ground to a halt as the Fraser River gold rush petered out, abruptly ending the business, building and population boom in Victoria, the key staging centre. In addition, there was the fact that the expected serious fires had not occurred, reducing the sense of urgency.

There was another factor in the decline of interest by the volunteer organizers as brought out by Laing and Lamb. It was the difference of opinion over control of the fire companies between Governor Douglas and those trying to set them up. On the one hand, there were the many men prominent in the embryo companies who had come from San

Volunteers of the Victoria fire companies assemble for inspection by Governor Kennedy in 1864. The squirrel-tail button engine, accompanied by torchboys is at left. Company banners, waving in the breeze, are all but invisible in the time exposure. - PABC-7814

Francisco and even had backgrounds as firemen there or in their hometowns in the east. They subscribed to the American way of independent fire companies who ran their own show. In addition to serving the community as firefighters, the fire companies were fraternal organizations or service clubs, with pride in performance on the fire scene, on parade with uniforms and regalia, and in competition. They did not want outside interference, especially from government officials.

Douglas no doubt was aware of the debit side of the system. More in the United States than in Canada, the highly independent volunteer fire companies swung strong political influence, feuded among themselves and sometimes pursued their other concerns at the expense of fighting fires.

The British administrator probably feared this aspect and may have been more familiar with the rowdier facet of the American city volunteer than with the more sober firemen of Canada's eastern volunteer companies. In Britain firefighting was being carried out by paid companies established by the fire insurance companies or by local authorities, often under the direction of the police. Following the British concept of direct government control, Douglas placed the fire apparatus under Magistrate Pemberton, with sole authority to act in matters of firefighting. Pemberton assumed the title of fire chief, but that seemed to be the end of it.

A parade of steamer equipped volunteer companies in Victoria receives a good turnout of citizens lined before the highly flammable wood-shingled buildings.
- PABC-8246

A grand jury looked at fire protection needs in Victoria in January 1859, expressing confidence that Pemberton would take the necessary steps for the appointment of fire wardens, engineers for the fire companies, the building of cisterns to supply the engines and for fire prevention regulations. The hint was taken to the point of two fire company captains being appointed with "power to organize companies and make use of the two engines belonging to the Hudson's Bay Company," as well as letting of contracts for the building of two cisterns, only one of which seems to have been completed. In June the chain gang filled it with 25,000 gallons of water.

But little was accomplished in getting the fire companies organized. There was no evidence of action towards the building of a fire station. It might be fairly safe to assume that there was a lack of enthusiasm on the part of those supporting the independent companies to join the government-sponsored units.

Editorial comment in the newspapers criticized the lack of progress, and after a warehouse fire seriously threatened the business district, *The Colonist* openly condemned the folly of putting the engines in the hands of the police instead of independent, enthusiastic volunteers with the right to choose their own members and officers. The only veto reserved by the government should be the right to dissolve a fire company for inefficiency, *The Colonist* asserted.

This sentiment must have been backed by popular opinion as, following the salutary experience of the fire, interested citizens quickly raised funds by subscription to purchase an alarm bell and apparatus for a hook-and-ladder company, without reference to the government. The Union Hook and Ladder Company was duly organized on November 22, 1859, a two-storey fire station erected at Bastion and Wharf streets, the bell installed, and a first class hook-and-ladder truck received from San Francisco. Again, the apparatus was secondhand, having been built in Sacramento in 1854.

Victoria's independent ladder company was in full operation early in 1860, organized on the model of American cities. This seemed to break the impasse, the independent formation of two engine companies being completed very soon, the 40 members of Deluge Engine Company No. 1 manning the Hunneman, while Tiger No. 2 took the Telegraph. The two companies occupied rented quarters around the corner from one another, which no doubt contributed to the normal rivalry in turning out for fires.

Edward Coker, who had been one of the police-appointed fire captains and now called himself Superintendent of Fire Engines, handed over the engines to the volunteer companies, and they commenced holding practices in March and April. This looks like at least a partial acceptance of the situation by the authority. The Deluge Company elected Governor Douglas a life member, then forwarded a request that the transfer of their engine be formalized and made permanent! Responding on a false alarm soon after they were organized, the Tigers had the misfortune to have two members trip and fall beneath the engine, suffering painful injuries.

By May 1, 1860, the fire companies of Victoria were organized to the state where they could parade through town drawing their decorated apparatus, complete with uniforms of red shirts, black pants and the glazed dress hats popular at the time. They were accompanied by the band of H.M.S. *Topaze*, torchboys and axemen. The Deluge carried an elaborate banner proclaiming its company name and motto. This May Day parade, instigated by the hook-and-ladder company, became an annual institution. Later in the day, the companies were reviewed by the governor.

The efforts of the companies were undoubtedly appreciated, but there was an evident fear of what uncontrollable political monsters they might grow into. There was, after all, the clear example of the problems they created in American cities like New York, San Francisco, Cincinnati and others. If the Victoria volunteers became obstreperous, the authorities could not afford to replace them with a paid force as the large American cities were doing. Thus, there was a standoff, the volunteer companies had the firefighters, but the governor controlled the fire engines (even though they still actually were the property of the Hudson's Bay Company).

An Act for the Protection of the Members of the Fire Companies of Victoria was passed by the House of Assembly in July 1860. While this was to protect them from legal liability, it recognized their existence and registered them as associations. The usefulness and popularity of the fire companies seemed to allay fears by officials to the point where the act was renewed the next year, giving them full recognition. The new version specified the form of government for the department as a board of delegates consisting of three members of each company. There were to be elected chief engineer, assistant engineer, president and other officers, all subject to approval by the governor. Now legal status had been given to an organization that was already in place.

Water supply then became a major concern, increasing numbers of cisterns or street tanks being constructed. In 1862 there were five of them, ranging from 15,000 to 32,000 gallons in capacity. Three more were added after this and some were in use until the mid-1880s.

Parsimonious Government

The fire companies of Victoria struggled along financially, depending upon sporadic government handouts, donations and fundraising events. The town paid nothing.

Late in 1862 the Tigers finally got a replacement for their aging and worn-out engine, the Telegraph, in the form of a bigger model by Button and Blake, prominent builders in New York State. Then the first real fire engine house was erected on Yates Street, between Broad and Douglas, costing $1,787. It housed the Deluge Company, which had been sharing in turn the quarters of the Tigers and the Union Hook and Ladder. Again, much of the cost of the new building was raised by public subscription. The next year two new hosereels and more leather hose were acquired, but the government was parsimonious. Then, as now, governments put a low priority on fire protection, except immediately following a disastrous fire.

Efforts to control the indiscriminate proliferation of flimsy, highly flammable wooden buildings were made in 1862 through an act establishing fire limits in the town of Victoria. The purpose of fire limits was to restrict allowable types of construction within the crowded high-value business districts of a city. They were an accepted part of building codes until the middle of the twentieth century, when the character of cities was changed by the automobile. In early Victoria, the fire limit area was laid down as that lying between the harbour, Johnson, Broad and Fort streets, a very small area. Only modest-sized, one-storey buildings were permitted within the fire limit. At this time Victoria became a city.

Toward the mid-1860s, the fire department's volunteer membership was waning. The work was arduous and it was expensive to be a member, but a major element must have been the feeling of being unappreciated by a niggling government. Only after the end of Governor Douglas' term of office was an exemption from most jury duties prescribed for the firemen. This helped restore the declining membership. There followed a somewhat muddled decade when the department's apparatus had been passed over to the city, but the fire chief was still reporting to the government. The chief engineer of 1865, J.C. Keenan, spoke out for improvements, but the decline of the Caribou gold rush was taking its toll on the economy of Victoria. The next year, the usual government grant of $3,000 to the fire department was omitted altogether and, in 1867, since Vancouver Island and British Columbia were now combined as one colony, Victoria had to split a $1,000 grant with the principal mainland town of New Westminster.

The firemen stuck to their jobs, however, sending a delegation to New Westminster to help celebrate the establishment of the fire companies there, repairing the Telegraph, and in a burst of effort the Tiger Company raised the necessary funds to purchase a steam fire engine. The Hudson's Bay Company, leading insurance companies and the government contributed towards the $4,005.40 cost. Even Sir James Douglas kicked in $20 personally. The first steamer in the West, it was a Button, by Button and Blake of Waterford, New York, and arrived via San Francisco in November 1868.

The men of the Deluge and Tiger companies assemble before the police station in Victoria. Apparatus is decorated for a parade, the Button steamer at left and an elaborate spider hose carriage at right. - PABC-57228

The latest thing in Victoria, Deluge Company's Merryweather. - PABC-57027

Not to be outdone, the Deluge Engine Company similarly raised the money to purchase themselves a steamer. They were sold on the lower-priced and allegedly superior Merryweather engine built in London, England. The volunteer companies were enthusiastic over their powerful new pumping engines. They do not seem to have seen them as a direct threat to the existence of the volunteer system as their American connections might have warned them. Perhaps they were reassured by the need for their musclepower to pull them to the fires, if not to do the pumping.

But the end was near. In the 1870s the city assumed control of approval for fire department regulations and elective appointments, the city limits were expanded, requiring the use of horses to pull the fire engines, and a waterworks system with hydrants was installed. The volunteers carried on, but on January 1, 1886, they were replaced by a force of 26 full-time firefighters.

The Royal City

The other major town in British Columbia at the time was New Westminster, the Royal City, named by Queen Victoria and designated capital of mainland British Columbia in 1859. Army engineers cleared the townsite that year at this location a few miles up the Fraser River from the sea. It was well located on a navigable waterway in the Fraser Delta, the one prime agricultural area, and on the route inland to the gold camps.

The new town elected a municipal council the following year. Like other new towns in the Canadian West, it was born at a time when public fire protection had become well established as a key service to be seen to by the authorities; not that there was any legal compulsion upon a municipality to do so, then or now.

A lot was set aside for a firehall, along with other civic buildings. In the summer of 1861, the locally famous Hyack Fire Company No. 1 was organized, Hyack meaning "quick" in Chinook.

With funds from the government and municipal council, a firehall and fire engine were provided. The fire engine house, Hyack Hall, was located on Columbia Street and had a large reservoir at the rear as a suction supply for the engine. The council established the New Westminster Fire Department, and F.A. Richards was elected first chief. As in many other cases the company included a number of leading citizens.

The first fire engine was purchased in San Francisco, a Button, at $2,600, with 500 feet of hose. To celebrate its arrival, a contingent of Victoria firemen escorted it over on the ferry steamer. The customary parades, band concerts and plenty to eat featured the new department's launching in the service of the young city.

The fire department included 58 men and five boys in 1863, and this was expanded the next year with the organization of the Hook and Ladder, Camp and Sapperton Company, with 34 members. More water supply tanks were also added at various points. The volunteers reigned until 1886 when the first career firefighters were added. As in many new towns hacked out of the forest, a great concern in early New Westminster was the surrounding bush, but the community was spared any major calamity from this source.

The greatest fire in the annals of the New Westminster Fire Department did not occur until 1895, on Saturday, September 10. At the wharf of Brackman and Ker, on Front Street near Lytton Square, there was a warehouse containing 200 tons of hay. In the age of horse-powered commerce, this was a vital fuel, needed even in the heart of a city. It was handled loose, not in bales, and was highly flammable. Arson was suspected when the hay erupted in flames at 11 p.m., and the hot fire very soon burned through the roof of the wharf shed. There was a dry, warm wind from the northeast to help the flames spread across to the north side of Front Street, into the exposed fronts of brick business blocks, and to the market building on the square.

If that was not serious enough, the flaming shed set fire to the steamer *Edgar* tied up alongside. The *Edgar* was cut loose to drift away, but the outgoing tide sent her against two

The commercial heart of New Westminister lies in ruins after the calamitous fire of September 10, 1895 that was spread by both wind and tide. - PABC-9367

sternwheelers, the three becoming floating bonfires drifting for 500 yards along the waterfront, setting fire to Cannery Wharf and various buildings along the docks. The Canadian Pacific Navigation Company shed was also full of hay, adding substantially to the conflagration.

From the burning line of waterfront buildings, the fire spread on the wind up the side streets into the business district, embers blowing far ahead. A rare fire it was, to spread in one direction on the wind and another on the tide. It crossed the 99-foot-wide Columbia Street, the main business thoroughfare, and onto First Avenue. Many businesses and the No. 1 Fire Hall were burned out. Chinatown was partly destroyed, everything went down as far as Royal Avenue.

The fire department at this time was equipped with one steam fire engine, a chemical engine, six full-time, six part-time and some call men. The chief was away at the time. As often happens in a fast-moving conflagration, hoselines had to be abandoned and the hydrants could not be shut down, with a resultant wastage of water and lowered pressures.

A large valve that could have been opened to let more water in from the higher levels was apparently overlooked.

There were many wooden buildings involved in this conflagration, but even the substantial brick business blocks on Columbia Street fell prey. The fire travelled across the street, through the blocks, parallel to the firewalls designed to protect each building from its adjoining neighbours. In about an hour a Vancouver Fire Department detachment of 20 men with two hosereels arrived under direction of their great Fire Chief J.H. Carlisle. They were able to save the Queen Hotel and another business block. In the light of the next morning the commercial heart and the waterfront of the Royal City was a scene of smoking desolation.

(XVI) Boomtowns

A Terminus for the C.P.R.

As the Canadian Pacific Railway stretched its rails across the near-empty breadth of Canada, the fate of townsites and would-be towns hung in the balance. The exact route of the first rail line was subject to modification, and whether it came through town meant life or death. This was very much the case in the selection of the West Coast terminus. Great decisions were required in selecting the most practical route through the enormous mountain range barriers to a suitable seaport on the Pacific Ocean.

Victoria, Esquimault, Nanaimo, New Westminster and Port Moody all had hopes. Port Moody was the railway's western construction base on the magnificent Burrard Inlet and seemed about to get the nod as the best site for a seaport. But by the beginning of 1886 it was definite that the winner would be the rough, scrappy, planked-street townsite of Granville, further down the same inlet. It was the final choice after some tough bargaining between the C.P.R. and the British Columbia government.

Early in the decade Granville was a lumbering site with about 300 loggers and squatters, between Coal Harbour and the Hastings sawmill, and an appendage to the mill site. Both were wedged between the tall primaeval forest and the sea. Across the wide harbour was Moody's sawmill, with comfortable little houses perched along the rocky shore.

Granville was a freewheeling, unkempt and unorganized settlement where lots sold for $100. Better known as Gastown, it was proclaimed a town on March 1, 1870, with the approximate boundaries Carrall, Cambie and Hastings streets and Burrard Inlet. Knowing its future was assured as the C.P.R. terminus, speculators swarmed in and Gastown began to boom.

Incorporation as a municipality was applied for and granted on April 6, 1886, as the City of Vancouver. It is said they had to pass the hat in the saloons to pay the expenses for a delegate to go to Victoria for the charter application. Growing explosively in a great outburst of anticipation, buildings were thrown up as fast as the thick forest could be cut down to make room. The C.P.R. also was busy clearing space for a roundhouse and terminal facilities.

A new city council set about making some order of things, and a meeting to organize a fire brigade was held on May 28. Going into June, the council, with no funds, was considering how to fund a fire engine. By now there were perhaps a thousand buildings cheek by jowl in the infant city, much of it a shantytown along the water's edge. On the boundary of the forest was a vast area of tree-cutting and great depths of slashings from the clearing operations.

On June 13 the city was two months old. The inexperienced city council, having no appreciable revenues, was struggling to secure some order in the chaos of mud, planked streets and crowded jerry-built buildings. A volunteer hose company and a hook-and-ladder company were organized under Sam Pedgriff as fire chief, but they had no equipment yet other than shovels, buckets and axes.

One of the first two fire engines in Vancouver, this Ontario-built Ronald at first drew water for fires from storage tanks.

John Howe Carlisle was elected volunteer chief by one vote shortly after the city was founded, but stayed on for 42 years at the head of a fast growing Vancouver Fire Department.

On this Sunday, C.P.R. workers were burning slash from tree-clearing operations at their locomotive roundhouse site. This proved ill-advised when the previously light breeze began to gust up to a gale during the morning, increasing the intensity of the burning. The slash fires spread beyond control into the surrounding accumulations of cut and drying branches. The great masses of tree trunks and dried-out evergreen branches produced by the hurried clearing operations were in heaps ten feet deep.

By early afternoon, smoke and embers from the now fierce fires were being driven across town by increasing winds. The citizens, becoming alarmed, began moving belongings from their buildings. But the explosive inferno roared down upon the town so fast that everyone had to flee for his life; great blasts of flame chased the inhabitants down the streets.

Many people took to log rafts and small boats, or leapt into the harbour or, in the opposite direction, False Creek. One hapless individual was overtaken and incinerated along with the horse and wagon he was driving along Carrall Street. There was complete panic and no time for firefighting, if there had been anything to fight with. The crowded, highly combustible buildings, the plank sidewalks and all the works of man that had been so eagerly put together to start the promising city were gone. Included among the structures lost were the city council chambers, the courthouse, two churches, six saloons, nine restaurants, 59 shops, 17 offices, 11 hotels, three livery stables and the newspaper office.

In 45 minutes the brave new city was reduced to a few acres of hot ashes. A handful of buildings remained, including two hotels and the C.P.R. station. The two sawmill sites, Hastings' Mills, and Moodyville across the inlet, were clear of the holocaust. (The earliest fire in the area was probably the burning of the earlier Moody mill in December 1873. It was rebuilt in a few months.)

A tent and shanty city quickly rose from the ashes, phoenix fashion. The lumber stocks of the Hastings' mill were made readily available so that makeshift shelters could be thrown together. Relief in the form of tents, clothing, food and blankets were sent in from New Westminster, Victoria and Moodyville (later to become North Vancouver). Mayor MacLean of Vancouver dispatched a telegram to Prime Minister John A. MacDonald, saying, "Our city is in ashes two thousand people homeless can you send us any government aid?" After a follow-up telegram, the prime minister replied that the Dominion Government would contribute $5,000.

Vancouver's fire department moved very rapidly from volunteers to fulltime manned stations as the city mushroomed. Early career men work out with their Hayes aerial ladder.

The young city council did not fail to learn something from the calamity. From their temporary city hall, made for them by Mr. Jones, the local tentmaker, the first financing was for the fire department — even before a jail or city hall. This was a harbinger of a long history of strong municipal support for the Vancouver Fire Department.

As with many other councils in their situation, it was necessary to rush to get tough building controls, since rebuilding was already going on at a feverish rate. Brick and stone structures arose, but there were still many sizeable wood-frame ones. Carrall Street had been quickly rebuilt in a few weeks, a range of crowded, shoddy firetraps.

Street tanks were installed for fire engine supply. Council ordered a modern steam fire engine, four Silsby hosereels and 2,500 feet of two-and-a-half-inch hose from the East. This made Vancouver unique among major Canadian cities, starting off with a steam-powered fire department instead of going through the stage of hand pumps. The engine was a Ronald, the very successful single-cylinder machine made in Brussels, Ontario.

The steamer arrived at the Port Moody railhead on July 26 and was hauled the remaining 12 miles to Vancouver by four horses. John D. Ronald himself arrived to test and demonstrate his product at the Cambie Street Wharf. The new fire engine was named M.A. MacLean, after the mayor. Like many of the citizenry, it lived under canvas for a time until a station was built.

The new fire engine and the firemen had their baptism of firefighting August 11, responding to a call to Spratt's fish oil refinery just outside of town. They got there in time to save adjoining exposed buildings, and there was considerable satisfaction with the performance. The press reported the engine "to be a superior article and the firemen, though inexperienced, worked with a will for an hour and a half and saved an adjacent warehouse and office." The citizens at large may have been happy to see the plant go, however, as the scowloads of fish waiting to be processed into fertilizer were reputed to stink worse than a glue factory.

One fire engine and station were not sufficient for the rapidly growing city, so a second Ronald was ordered and received the following May, a benefit ball being held by the volunteers to raise funds. Indicative of the rapid spread of this boomtown is the fact that a charter for a street railway was taken out in only the fourth year of its existence.

Later on in 1886 an election was held to choose a new fire chief, and it was decided by one vote. The winner, John Howe Carlisle, deserved a stronger endorsement. He remained in office to become the first professional chief of the department and to guide it through enormous expansion and developing technology over a period of 42 years. He had sagacity and the progressive approach of a keen student of his craft.

By 1889 the fire department occupied two firehalls with a full-time force of 15 career firefighters, a Hayes aerial ladder truck, and horses to pull the heavier apparatus. The superb Capilano water supply from across the inlet was in service in June, providing hydrant service. Modern alarm communications, in the form of 15 street fire alarm telegraph boxes and telephones in the firehalls, were added also.

The volunteer hose companies attached to the two stations were Invincible (No. 1) and Reliance (No.2). Their prowess was recognized when they became North American champions in the hosereel race, beating 12 American teams in competition in Tacoma, Washington, in 1889, including Rochester, New York, the longtime titleholders.

Gold Rush Country

Dawson, the Yukon gold rush capital, sprang up almost overnight. The news of rich gold deposits in the nearby creeks in 1896 brought the miners and would-be miners rushing from all over the continent and even overseas. By the autumn of 1897 the collection of shacks along the river had become the ultimate boomtown. Along with slogging for gold in their diggings, the miners lived it up in what has been described as "a maelstrom of uproarious enjoyment." The Opera House and the M & M were always packed to the doors with revellers.

But Dawson was different from the lawless mining towns of California, since the Royal Northwest Mounted Police were on the scene early and quietly kept things under control (without the dramatics of a Wyatt Earp). The American crime syndicates that liked to take over whole towns were turned away before they could become established.

On the fire front, however, Dawson was, from the beginning, a conflagration looking for a place to start. It was a conglomeration of hurriedly thrown together wooden buildings, shanties and hundreds of tents crowded on a narrow flat along the Yukon River. The time and place came together on Thanksgiving night, 1897. The M & M, on which owner Pete McDonald had just paid off his final debt that evening, caught fire, said to have been set off by one of the dancehall girls upstairs.

A bucket brigade of bystanders lugging water from the nearby river had little effect on the advancing flames. The crowd then turned to demolition of the building and especially to the rescue of the large stock of liquor. The fire was rapidly out of control, engulfing the adjoining Opera House, then the Dominion, another entertainment centre, gutting the heart of the town.

This first conflagration pointed up the obvious hazard inherent in the jerry-built town and the critical need for an organized and equipped firefighting organization. A meeting of businessmen on March 10, 1898, raised subscription promises of $20,000. Five days later fire apparatus was ordered, including an Ahrens steam fire engine and apparently hand-drawn two-wheeled reels. It all arrived by steamer from Seattle on July 16, but the promised subscriptions had not been forthcoming to pay for the order. It languished in storage, therefore, the steamer partly assembled and in a crate. It seems the erstwhile subscribers wanted the government to pay for it.

This was the situation on October 14, when at a few minutes after 6 a.m., fire was discovered in the Greentree Building, allegedly started by the same dancehall girl as before. The police sounded the alarm by bugle call, but the building was soon a seething mass of flames. The post office followed suit within 20 minutes, and then the magnificent $100,000 Worden Hotel succumbed, the occupants barely having time to escape.

The fire spread in three directions, devouring 16 businesses and about 20 "cabins" or homes on Second Avenue. The whole city was threatened. Fully 2,000 men were on the scene, but with scanty firefighting tools. One chemical engine was available, supported by hundreds of men with buckets, basins and whatever other utensils they could find, all of little avail. After several hours of retreating from the advancing heat front, the crowd was amazed and cheered to see a large-as-life steam fire engine put in an appearance and head for the river.

It was none other than the Ahrens engine, hurriedly unpacked and assembled without benefit of paid invoices. The incredibly rapid assembly job was performed by employees of the N.A.T. & T. Company in two hours, under the direction of a former engineer from London, England, and helped by several former firemen from other cities. A line of hose was quickly run out to the river and a strong hosestream brought to bear to save the day, according to contemporary accounts.

Organization of a fire department quickly followed. In answer to an advertisement published in the Dawson *Daily Dagger*, a meeting on October 19 set up a Board of Fire Commissioners, 100 firemen signed up, and a chief, assistant chief and three captains were elected.

There was no water supply other than the Yukon River, running parallel and close to the main street, First Avenue, and this was frozen solid to a considerable depth for many months of the year. The solution for the fire department was to station the fire engine on the ice under a tent, with its suction hose lowered down through ten feet of ice to the running water below. Steam was kept up constantly by several stokers hired for the purpose. In case of fire, hose was run out from the steamer to the shore and inland to the fire scene using the hand-drawn hose jumpers.

Where shall we sleep tonight? Dawson, like the earlier pioneer towns faced severe shortages after a conflagration, even to boards, nails and tools to rebuild. - PAC-C53170

Dawson firefighters of the Bonanza company, a mix of adventurers from far flung origins, pose before their hose house. The great risk of fire spread between buildings is evident. - Kinsey and Kinsey, PAC-C8145

What is rated as the "Great Fire" among the several conflagrations in Dawson occurred in the winter of 1899. It started in the Monte Carlo, the biggest building, located in the heart of town, and spread with great speed among the crowded buildings. There had been an altercation over pay to the stokers of the steamer and there was no steam up for pumping. Kerosene, sawdust and wood were plied to the boiler in a frantic attempt to get the pump going for a hoseline, but it was too late.

In the short space of two hours a major part of First Avenue was flattened. There was no wind, but the buildings were tinder-dry inside from the high heating by stoves all winter. The outside temperature was -45 degrees Fahrenheit and the hose froze solid after brief use. Blowing up buildings served no advantage, as has almost always been the case. A large part of the City of Dawson was again laid waste.

Dawson City had sprung up in the wilderness of the Yukon Territory in a spectacular fashion, as thousands of hopefuls responded to the lure of gold lying about in nuggets waiting to be picked up. The site was far from civilization, almost everything having to be imported from far south, usually by ship along the coast from Seattle and Vancouver, then up the long Klondike River during the short ice-free summer. The cost of consumer goods was very high, when available at all. When fire destroyed a building and its contents during the seven months of winter, there was a long wait for replacement of building materials, food and furnishings. A sawmill had been established to supply lumber, but when fire destroyed it there was a great shortage, forcing a return to the primitive system of whipsawing logs by hand. Nails were also said to be in very short supply after one fire, such that they were being reclaimed from used boxes, crates and any other available source.

The fire department was beefed up in 1899, apparently with territorial government financing. A second steam fire engine was delivered in the late summer. The department was now under the leadership of Acting Chief H.A. Stewart, only 26 years old, a native of Georgetown, Prince Edward Island, and a former mill owner. Stewart maintained one engine under steam at the river and a hosecart ready to lay out hoselines from it at all times. The second engine responded as needed from the firehall.

A tough consolidated fire regulation was brought out in June 1900 under the territorial commissioner, covering Dawson and all the Yukon. A full-time fire inspector, L.N.B. Bullock, was appointed to enforce it. It was a fire safety regulation typical of many cities of the day. It tried to restrict the use of highly combustible materials such as hay, straw, wood shavings, kerosene and gunpowder, as well as sources of ignition from smoking, ashes and stoves. The inspector closely watched the universally used stoves and the long metal stovepipes for proper clearances from woodwork and deposits of flammable creosote.

The fire law vested regulatory powers in the fire chief, and through him his subordinates, and was essentially the same in approach as the provincial fire commissioners' and fire marshals' prevention acts that were just beginning to be developed to the south. In Dawson, all two-storey buildings required permanent side and roof ladders, useful both for escape and for firefighting access. Makeshift housing was a hazard the authorities were now trying to eliminate. In the early days there were acres of tents, as men flooded into town. These were eventually forbidden, along with buildings utilizing straw, hay, manure, moss and sawdust anywhere near the built-up part of town.

The central business district of Dawson at that time was composed of eight blocks of large 1½- to 2½-storey wooden buildings packed closely together, there being only one brick structure in the district. Bullock said the locally-made brick was of poor quality and chimneys had to be hung on the building, hence the use of iron stovepipes.

By 1902 the kerosene lamp and wax candle were being supplanted by electric light and new roofs were covered in sheet metal. A strong fire department was now in place under 28-year-old Chief Stewart. He had two firehalls, three steam fire engines, a big Waterous engine having been added to the Ahrens and Clapp and Jones engines in 1900. These were supported by two Champion soda-acid chemical engines, a Seagrave ladder truck, and various hosereels and wagons.

It was now a horse-drawn fire department manned by 19 full-time career firefighters. Typical of the Dawson of that era, these men had arrived in town from widely scattered places, all within the past five years. Most were bachelors and only one was over 40 years of age. Two of the steamers were maintained on barges in the Yukon River, ready for instant duty, the installation of a piped waterworks system and hydrants being considered totally impractical. The barges floated at their moorings in the summer and froze solid in the ice in winter. The third engine ran from a firehall until 1901, when a well was dug on First Avenue and a building erected over it to house the engine. A modern fire alarm telegraph system was installed, having 19 call boxes located on the streets, the equal in modern efficiency of the major cities in southern Canada.

To protect its population of somewhere between 4,000 and 6,000 inhabitants, in 1902, Dawson had a strong fire department paid for by the territorial government. This free ride was soon to end, however, as the city was to be incorporated and become responsible for its own protection. It would bring about a return to volunteer manpower in the end, as the city eventually declined.

The hastily-built gold mining towns of the Cariboo also made for easy burning. Organization of any kind, including fire protection, was haphazard. Commercial and residential buildings of wood were jammed eave to eave along a narrow main street, often a narrow defile between steep hillsides. The goldseekers wanted only the basic necessities, including saloons, of course, and did not plan on staying long. Barkerville was one such town. Some of these towns led miraculous lives, but Barkerville was wiped out in about three-quarters of an hour, from end to end, in 1868. It was promptly rebuilt with government help.

Desolation quite abruptly replaced homes and thriving businesses, taking only a couple of hours in frequent conflagrations a century ago, in this case St. Hyacinthe, Quebec, in 1876.

5

BIGGER CITIES, BIGGER FIRES (1880-1920)
(XVII) More Painful Lessons

In the last quarter of the nineteenth century, Canadian cities reached a stage of fairly sophisticated fire protection. They had waterworks hydrant systems, steam fire engines, elaborate horse-drawn apparatus, total or partial career fire forces and often some semblance of control over building hazards. Still, some notably disastrous fires made cities reel in shock. Increasing insurance coverage helped, but did not negate the losses. The chief problem was that the growing complexity of buildings, commerce and comforts was outpacing the growth in competence of the fire control system. Keeping building codes abreast of new ideas in the design of buildings was to become a perennial problem, still not solved.

In the 1870s there were still many large fires of conflagration stature in Canadian and American cities, even though solid brick and stone-walled structures had taken over the more crowded commercial districts. Though built with incombustible walls, increasing heights and areas produced bigger fuel loads and fire intensities, calling for new measures of control to be built in. There was the need to subdivide large areas by compartmentation, by building fire resistance between floors and giving protection against radiated heat through the windows. The great boon of the automatic sprinkler system still lay some years off.

The vulnerable business and industrial concentrations intensified, growing upward more than outward. Local transport was still by horse and wagon, dictating location as close as possible to other buildings, the railway and water transport. Factories and warehouses were all multi-storey, the invention of the elevator making it efficient. It was only after World War II that this pattern was broken, commercial building spreading out into cheaper outlying land and rising only one storey. The era of the horse and wagon and the elevator was succeeded by that of motor transport and forklift trucks.

The ensuing spread and need for vehicle yard areas made for generous space between buildings. The building codes also offered the alternatives of space or structural protection to inhibit fire spread between structures. The automatic sprinkler had come on the scene in the 1870s and was perfected toward the end of the century. It was a response to the extreme hazard of the big lint-filled cotton mills of New England, but found a universal application in abating the fire risk in all manner of building. A great ally for the fire department, the automatic sprinkler system cut to a manageable level the risk of the large or hazardous property. In very recent times its great value in ensuring life safety has come to the fore.

Devastation in Smaller Cities

Not many of the old eastern cities and towns escaped having a calamitous fire at some time in their histories. There were always conflagrations occurring in the smaller cities in Canada, too numerous to mention even in this book, during the latter nineteenth and early twentieth centuries. Weak building controls, resulting in crowded wooden buildings or flammable back sheds of brick ones, and the prevalent wood-shingle roofs were key elements. Ineffectual fire departments and insufficient or disrupted water supplies usually contributed. Strong winds and dry weather frequently helped the fires get their essential headstart. Similar conditions and similar results were equally in evidence on the American side of the border.

St. Hyacinthe and St. John's (St. Jean), Quebec, nearby towns south of Montreal, both had severe fires in 1876. The former lost 800 buildings in an area covering 100 acres, including the entire business district. The waterworks broke down and the fire department was both inefficient and poorly equipped. Help was rendered by the fire departments of Montreal, Quebec and Levis. St. John's had a fire rip through the length of its two principal streets, for a distance of a mile, taking out the principal business establishments, 200 stores and dwellings, the woollen mill and docks, a $3,500,000 loss.

St. Hyacinthe suffered another major conflagration on May 20, 1903. This one started in a boot and shoe factory of four storeys. On a freshening breeze, the fire spread to destroy every building in 32 blocks. (This town would have two more conflagrations, the latest in 1981.)

It was Windsor, N.S., in 1897, Sydney in 1901, Trois Rivieres in 1908, and Campbellton, N.B., in 1910. The last two cities were largely destroyed. In Campbellton, only seven houses escaped, while most of the village of Richardsville, a mile away, was also burned. The Trois Rivieres Fire Department and water supply were weak, and officials, lulled into complacency by a good record, had ignored repeated warnings of the danger to their city, one of the oldest in the country.

Other calamitous fires struck cities and towns hard, not by burning out a large number of buildings, but through destruction of a major or only industry. Such a fire, and one that was of special interest to fire protection engineers because of its effect upon fire-resistive construction, occurred in Peterborough, Ontario.

Over a period of years, beginning in 1901, the Quaker Oats Company, cereal manufacturers, built a huge complex of buildings ranging four to nine storeys in Peterborough. The complex was composed of substantial warehouses, storage elevators, and manufacturing buildings, interconnected and stretching along the Otonabee River. Most of the buildings were protected by automatic sprinklers. On December 11, 1916, a violent explosion, the one enemy that can cripple an automatic sprinkler system, blew out two walls of the dry house, killing 22 of the 30 to 40 workers in the building. Many rescues and other deeds of heroism were performed by those lucky enough to escape.

Flour and cereal mills are subject to rapid fire spread or explosions if stringent dust control is not adhered to. In this case, it took only a couple of minutes for the fire following the explosion to rush into the adjoining flour mill at all floor levels. The explosion wrecked sprinkler pipes and the mill fire pump steam supply. With poor hydrant distribution in the vicinity, the Peterborough Fire Department, a chief and only 16 men, faced an overwhelming task. The fire spread steadily through the mill into two big warehouses. Even though much of the construction was fire-resistive reinforced concrete, the high fire load in the contents created a fire of severe heat. Fire experts were fascinated with the oven effect of such great heat that melted bricks so they flowed like molasses. The reinforced concrete spalled and dehydrated so that columns and floors sagged or failed.

This fire and others like it established that fire resistance is a matter of degree and directly related to the fire load of the structure's contents.

While it did not apply in the foregoing case, the loss of a major or only industry has often been calamitous for a town. It can mean anything from lost wages, until a plant is rebuilt, to the demise of the industry. A burned-out plant may be the last straw that kills the company or the incentive to relocate elsewhere because of changing market conditions.

It took dramatic cases like the Quaker Oats fire of 1916 in Peterborough, Ontario, to demonstrate clerly that even solid noncombustible structures were not impervious to fire without protective design features. PAC-C32921

An enormous conflagration rolls through two cities that refused to foresee disaster. The scene on Queen Street in Ottawa as flames approach, April 26, 1900.
- Booth Family, PAC-C120334

Ottawa — Hull Spectacular

In the late nineteenth century, the Ottawa area was the greatest lumber-producing centre in Eastern Canada, shipping Ottawa Valley and Gatineau River pine in great quantities to the United Kingdom and the United States. Mills to process timber for the North American markets crowded around the Chaudiere Rapids of the Ottawa because they were the source of power, the foot of the upper river log drives and the head of river navigation to the St. Lawrence River and the sea.

The twin cities of Ottawa and Hull did not build ships to load with deals, selling both on arrival overseas as did Saint John, but thrived then on producing finished lumber. Secondary industries included sash and door, wooden buckets, furniture, matches and sulphite pulp for paper. Hydro-electric power from the Chaudiere also lighted the cities and ran the street railway, while water motors supplied pressure for the water mains.

Like Saint John before, the cities facing each other across the river were in large part built of their own product, virtually all of Hull and the whole west end of Ottawa. It was an enormous accumulation of wood. There were sawmills, bunkhouses, workers' houses and tenements and stored lumber, all cheek by jowl. It was light frame construction with wood shingles, and great stacks of finished lumber as high as a four-storey building piled on the wharves along the river and in every available space throughout the west end. Much of the lumber was piled with air spaces through it to aid in seasoning, but also increasing burnability.

Since everybody walked to work, homes were crowded near the hub of industry, the low-lying area on the river around the Chaudiere Rapids, Rochesterville, Victoria, and across LeBreton Flats and into Hull. Even the great homes of the timber tycoons and managers were here. The big names — Eddy, Booth and Bronson — were entrepreneurs in timber, manufacturing and transportation.

The fire insurance companies were certainly aware of the fire risks of this hive of prosperity, as well as the weakness of the Hull and Ottawa fire departments. There were grumblings in Ottawa over high insurance premiums in the 1890s. Long before, Alfred Perry, the veteran firefighter and chief inspector for the Royal Insurance Company, had made a specific prediction and given a warning to the city officials. In his reports he had said that someday Ottawa would be visited by a very destructive fire. It would start in Hull, he had surmised, cross into the Chaudiere district on the river flats, extend up into Rochesterville, aided by the continuous piles of lumber, and, if the wind was in the right direction, the best part of the city would be swept. Two barriers to this possibility were the 1,000-foot open space of the Ottawa River and the bluff that separated the industrial flats from the higher level of downtown Ottawa.

When it came to passing bylaws to regulate construction, the Ottawa city council was easily intimidated. Great influence was wielded by the industrial barons, who could threaten to move out of the city, taking their tax money with them. J.R. Booth had already once moved out of the Ottawa limits, only to be captured again with his payroll of 2,000 employees, some $5,000 a day, including taxes, as Ottawa's boundaries were extended westward along the river. Hull was the poor neighbour with little hope or inclination to attempt to control fire-hazardous building.

The hazard was ripe in both cities and across the industrial link between them, on the riverbanks and islands. Ottawa had only two fire engines, Hull none. In this setting, then, true to the fire underwriters' code, conflagration weather brought the final key ingredient for the conflagration recipe. On Thursday, April 26, 1900, it was a balmy Spring day at 63 degrees Fahrenheit, but with a dry north wind at about 30 miles per hour. At about 10:30 a.m. someone stoked the fire in a small frame working-class home in an area of congested frame dwellings in the north end of Hull. There are two versions of the cause of the fire. We can choose a spark from the chimney or an overheated stovepipe. It does not matter, the sun-dried wood-shingle roof was soon ablaze, and another house and a barn were involved by the time the fire department arrived with its hosereels. The near-gale wind carried flaming brands through the tinderbox neighbourhood. This kind of conflagration, travelling from rooftop to rooftop, may be likened to a crown fire in a forest, moving with speed across the tops, leaving what is underneath to burn later.

Any efforts by the Hull Fire Department, with its small force and four hosereels, were of no consequence. Almost all the buildings in the city were of wood-frame construction and sheathed in cedar shingles. The storm of flame spread unhampered through the southern and western half of the city, including the industrial strip on the bank of the Ottawa River, which it reached in about two hours, having travelled eight or ten blocks.

The conflagration wind increased. By noon, all the houses on many of the principal streets, including the post office, Customs House, Imperial Hotel, the Anglican church, 15 stores in the business district, and the fire station had succumbed. Driven on the wind, the rain of burning material in the air bore down on the E.B. Eddy mills and the wood piles of the Hull Lumber Company. Mr. Eddy likened it to a "snowstorm of particles of fire." He said the wind drove sparks up underneath the shingles right inside his warehouse. The flames roared through the airspaces in the building-sized lumber stacks on the wharves and on the shore. Wooden-trussed flumes, that floated logs from the river, now carried fire.

A giant pall of smoke loomed across the river, and from it came the rain of fiery brands. The river was not insignificant, 1,000 feet wide. Although the steam fire engine Union, operated by the mill owners, and the Ottawa Fire Department's Conqueror and LaFrance engines fought valiantly to stem the crossing, fires started all around them on the south bank. The wooden deck of the bridge linking the two cities served as a fuse to carry the fire as well.

There were thousands of workmen on hand to help in trying to save their places of work, but there was no stopping the blizzard of incendiary particles landing in the various lumber yards of the Booth and Bronson companies in Ottawa's Victoria ward. The workers, in many cases, soon gave up and ran home to see to their families and personal property.

By two o'clock the fire was gaining a hold in the Chaudiere district, destroying the towering structure of the McKay flour mills and the Baldwin Iron Works, as well as the hydro-electric plants of the street railway and electric lighting system of the City of Ottawa. The important industrial district was rapidly swept through, the firemen having to abandon the Conqueror to be vanquished by the flames. All the facilities of the Canadian Pacific Railway, including the railway station, were lost.

The two city fire departments were totally overwhelmed by the speed and wide front of the fire's advance. It likely made little difference that these two fire departments were unreasonably weak to begin with, once the conflagration was rolling. Nevertheless, the Ottawa firemen managed to save the city waterworks pumphouse, source of supply both for domestic and firefighting use.

Help was requested from Montreal, Toronto, Brockville, Smith's Falls and Peterborough. The Montreal Fire Department rushed an engine, hosereel and eight men by extra train on the Canadian Atlantic Railway, making the trip in just under two hours. Another M.F.D. engine arrived in the evening.

Having eaten out the industrial heart of the two cities, the onrushing conflagration ran on southward through the Rochesterville residential district, right to the city limits at what is now Carling Avenue, in the vicinity of the St. Louis Dam at Dow's Lake. With some justice, perhaps, the palatial home of timber baron John Booth was erased along with the modest homes of his employees.

A defense line was set up by firemen and three militia companies using buckets, under command of Sam Hughes, along Booth Street, but it was mainly the cliffs of the escarpment that kept the fire from working eastward into central Ottawa. As was frequently the case, the use of dynamite on frame buildings to clear a firebreak aided more than it hindered the fire's passage. Also, as in many conflagrations, occasional buildings were spared when all their neighbours were levelled. A wooden building unaccountably untouched when adjacent solid masonry ones were burned out was a familiar scene.

On the other hand, several solidly built industrial buildings were saved by their superior construction. Notable was the survival of the large Booth sawmill, likely built of wood, but protected by an inside-and-out sprinkler drencher system designed by the owner's son and fed by the company's own pumps. The employees helped to save the mill, then dashed to their own homes.

The fire demon was satiated by midnight and the view of the blazing ruins from the higher levels after dark was described as awesome in its beauty, the ugliness and evidence of disaster masked by the night.

The size of the area destroyed and the number of homes and jobs gone were enormous, the dollar loss comparable with the much smaller area conflagration occurring a little later in the heart of downtown Toronto. Finished lumber stored in a long arc of company yards from Hull to Dow's Lake, one hundred million board feet, was a $3,000,000 loss. There were 1,400 buildings burned in Ottawa, dispossessing 14 percent of the city's 58,193 population, covering over 440 acres and representing an equal loss. In Hull 1,300 buildings were burned, throwing 42 percent of the 13,887 inhabitants on the street. The loss on that side of the river was $3,300,000, the burned out area being 276 acres. Seven lives were lost in the fire.

The Booth lumbering empire lost most of the finished product in its five yards, but saved the main mill; Eddy's plants all went down except for the sulphite mill; the hydroelectric power supply was gone; and thousands were homeless and out of work, at least temporarily. The Ottawa authorities, acting decisively for a change, quickly organized a massive relief programme. Everyone had a roof over his head within 48 hours. Large-scale relief came from far and wide, including food, furniture and a million dollars in cash. It would appear to be one of the last cases in Canada in which other cities rushed special trainloads of food, furniture, clothing and building materials to the aid of another in need. In future it would be cash, usually from higher levels of government, or funds from public donations.

The insurance companies paid their claims, providing at least the business firms the wherewithall to start rebuilding. But when reconstruction of the burned districts repeated the same hazardous situation as before, they surcharged the rates heavily, for the ineffectual Ottawa council was again bullied by the big entrepreneurs and the Board of Trade into allowing these conditions. Control over the indiscriminate piling of lumber was overruled, the council evidently blackmailed by the threat to rebuild the industries outside of the city.

It would also have been a hardship to demand that the poor workers build brick houses with lumber so cheap, so the city settled for a requirement of noncombustible roof coverings and 60-foot spacing for lumber yards. The big mills and their lumber piles eventually moved to other locations for commercial reasons anyway, and with times getting tighter, small companies were absorbed by the bigger ones. The fire accelerated a shakeout that eliminated the weak ones.

In the interim, the lumber piles were rapidly recreated by the rebuilt mills and the surviving Booth mill. After three years of rebuilding, on May 11, 1903, a fine Spring morning with a fresh southerly breeze, the Ottawa firemen had a fire at the Broad Street lumber yards. They dashed to the scene and promptly extinguished it. The fire companies had just picked up and returned to quarters when the call came in signalling a rapidly growing cloud of smoke rising from the lumberyards at the Sparks estate west of Preston Street, the precise point of origin not being recorded. It was rapidly developing into a major fire, this time pushed by a south wind, carrying flaming brands in the opposite direction through Rochesterville, from a point approximately where the 1900 fire had ended. This time it was not wood shingles but lack of water, high winds and closely exposed wooden homes that were to blame, the roofs now being of composition or sheet metal.

At the height of this fire, one of the two 24-inch-diameter principal supply mains leading out of the Ottawa Queen Street waterworks pumping station burst, flooding the station to a depth of five feet and severely reducing the pressure in the system. Fortunately, by heroic efforts, the waterworks employees were able to isolate the break and keep the plant running, getting a reduced supply of water to the firefighters. The main had been under repair and failed under increased pressure, it being the practice to raise the pumping pressure when there was a fire, a common procedure then in many cities. Eventually it was realized that this tempted fate, resulting in cases like this one. If a main was likely to break, it occurred at the worst moment — during a fire under the stepped up pressure.

Aided by the interruption in water supply, the flames ate their way north along Rochester and Division Streets as far as Albert, then were stopped again from entering downtown Ottawa by the rocky escarpment. The entire Chaudiere sector "fell under the menacing hand of King Fire," as it was put. Firemen, militia and citizens worked feverishly to save individual properties, sometimes successfully. Once again, the Montreal Fire Department rushed two hose-and-engine companies with 20 men by a C.P.R. extra.

This second conflagration did not get so far, burning out 34 blocks, a mere 75 acres, representing a loss of 300 homes and 18 million board feet of lumber, the monetary loss about $1,000,000.

With the two fire disasters there was finally sufficient backing for the Ottawa civic authorities to impose more severe limits on the piling of lumber. In any case, the fortunes of the lumber industry in Ottawa were now fading as the market declined and the mills moved out to be closer to new stands of timber. Soon this primary industry of Ottawa would be replaced, during World War I, by a burgeoning and possibly indestructible new industry, federal bureaucracy.

Lessons in Structural Protection

Two more great conflagrations were about to occur in the downtown hearts of two major Canadian cities, the last on such a scale in this country. They would precede the end of the horse-drawn and steam pumping fire departments. Slowly improving building codes, a better understanding of the difference between noncombustibility and fire resistance in construction practice, and the growth of more powerful fire departments utilizing gasoline-engined fire apparatus would then have a telling effect on the record of communities' vulnerability to fire.

At least this would be so in the congested commercial cores of cities, where it was considered appropriate to concentrate the most stringent standards of structural protection, one building from another. Wood exterior walls were forbidden within prescribed "fire limits." Many wood-shingle roofs caused conflagrations in New England and Canada, forcing them to be largely outlawed in the old eastern seaboard cities. Today the western half of the continent is just going through the process of catching on to this risk, now appearing in the form of cedar shake shingles, notably in the very dry areas of the southwestern states.

The better understanding of structural fire protection came in particular from careful evaluation of a number of North American conflagrations by increasingly alarmed fire insurance companies, now banded together in the National Board of Fire Underwriters to perform such studies on an engineering basis. Buildings thought to be fireproof were effectively wrecked if not destroyed, it was found, because, although the structure did not burn, the heat of the burning contents often damaged them severely or caused their collapse. This heat also was sufficient to spread fire from building to building by radiation through adjacent wood-framed, plain glass windows across alleys or streets.

The classic case was the fire of 1904 in Baltimore, Maryland, involving 80 downtown city blocks, in which many well-built fire-resistant buildings were burned out but left standing structurally intact. San Francisco was a similar case; a number of fires caused by the earthquake of 1906 spread through the entire business district of substantial buildings. Some large office buildings in use in San Francisco today withstood the earthquake but had to have their interiors rebuilt after the fire swept through them.

Modern building codes have done much to eliminate the weaknesses that allowed fire to race through buildings unhindered, but a considerable legacy of old construction is still with us.

New insurance and building codes were written to call for the protection of openings, both horizontal and vertical. Now, instead of the iron shutters of old, it was wired glass in metal frames for windows and furnace-tested fire-resistant door designs. The new buildings, their steelwork covered in masonry, their windows protected from their neighbours; and floors and vertical shafts protected against vertical firespread, were now more cautiously labelled "fire resistive" buildings. The term "fireproof" has long been abandoned by fire protection professionals.

The point was driven home by one notable single-building fire in 1912. The eight-storey, block-long headquarters of the Equitable Life Assurance Company in New York was built with substantial fire resistance, but vertical shafts and other concealed spaces allowed the hot gases of the burning contents to spread inexorably through it from end to end. It was a rude awakening to have this happen in spite of the all-out efforts of the world's most powerful fire department.

The two great Canadian conflagrations occurred in Montreal and Toronto at the beginning of the new century.

The Board of Trade Conflagration

Since the change to a full-time or career fire force in the 1860s, Montreal suffered only one more major conflagration in its central business district. This is known as the Board of Trade fire and occurred in the winter of 1901. While the fire department was not particularly strong in the latter part of the century, it was professional early and had the decided advantage of being backed by a water supply far stronger than those of most cities. From the 1850s on, direct hydrant streams were the rule and hand-engines and steamers were not depended upon.

In addition, mercantile construction was substantial, with heavy masonry walls, frequent metal window shutters and metal roof sheathing instead of the wood shingles of other cities and the outlying districts of Montreal itself. This seems to explain why the narrow streets of the old city did not breed more large-scale fires. Nevertheless, the insurance underwriters were uneasy and complained of weakness in the fire department as the nineteenth century closed. The buildings were growing, and the fire department was growing with the expanding city, but the once strong water supply was becoming relatively weaker.

What the well dressed fireman wore at the turn of the century – a rubberized fabric coat and tall leather helmet.

Plain glass windows and plentiful wooden trim on otherwise brick and stone buildings allowed flames to beat the fire department in the Board of Trade Building conflagration of 1901 in the heart of old Montreal. - PAC-PA28946

The Board of Trade Building was the Place Ville Marie of its day, a marble-lined palace of business, six storeys high and occupying an entire city block in the heart of the wholesale district. It housed 150 tenants, including many prominent business offices. It was widely considered to be fireproof and was divided into three sections, but had not been designed to be fully fireproof, even as this was understood then. It had a steel frame, but this was unprotected against heat.

The fire broke out at 7:30 p.m., January 24, when the premises of M. Saxe and Sons' clothing factory and warehouse, on narrow St. Peter Street at Lemoine, erupted in flames. The fire in the highly combustible interior trim and contents burned so fast that the first-due fire companies had only had their hoselines in action about ten minutes when the front wall collapsed into the street. Two employees had been working late and were lost in the flames.

Nelson's fancy goods warehouse across Lemoine Street ignited immediately from the flare of radiation. No high wind was reported, and the explosive spread in many directions among the fuel-packed buildings seems the only explanation for the ensuing conflagration. As was stated at the time, whenever the fire lunged in a new direction, the firefighters seemed to be fully engaged elsewhere. Across St. Peter Street from Nelson's, the Board of Trade's cornices and window trim on the upper floors began to smoke. The firemen might have handled this, but the protective blank wall of the Nelson Building collapsed against the huge office building, about 30 feet distant, and its doom was sealed.

The firefighters were stretched on many fronts. Every piece of fire apparatus in the city was called in. Operations were under the command of Sub-chief Jackson, Chief Benoit being ill. The firefighters managed to keep the fire from spreading westward, but it pushed them relentlessly back on the east side. Reluctantly retreating, they backed through a full city block to St. Nicholas Street and then south through to the fur trade businesses of old St. Paul Street, across that narrow thoroughfare and through another block to Commissioners Street.

The great 120-foot width of this street stopped the marching flames about midnight, which saved the waterfront adjoining. Here was another case like Saint John in 1877 and New Westminster in 1898. Buildings had blank walls and roof parapets as protection from their neighbours alongside, but the big fires crossed streets and moved through blocks by way of the unprotected front and rear windows.

Huge crowds, unruly and reckless, gathered to jam the streets of Old Montreal, sleighs and cabs obstructed the firefighters. Help came from the suburban fire departments of St. Henri, St. Cunegonde, Maisonneuve, Westmount and others. Ottawa was asked to send a steamer, but declined since they had only one following the loss of the Conqueror.

At 10:25 p.m. the Board of Trade was what a newspaper described as "a fiery mass spreading in a glow of fervent heat." Embers flew as far as the slopes of distant Mount Royal and rained down on the uptown business area as far as Sherbrooke Street, a mile away. A good snow cover prevented a much greater disaster. No firemen were lost, though there were many close shaves as they fought stubbornly in the narrow streets, retreating along St. Paul Street and operating their hosestreams from the shelter of doorways. The watertower was set up, but no steamers were used, hydrant pressure being relied upon.

The Board of Trade fire caused a sudden awakening from the complacency of a good record, a familiar story. There was much discussion of what went wrong and a formal investigation and defense by the Fire Commission without any conclusive findings. In his report for 1900, which was written after the fire, Chief Benoit made some remarks of a defensive nature. The insurance companies, who had been uneasy before and had to pay out $2,026,000, immediately implemented a whopping 50-percent increase in premiums. Commenting on the calamity which gutted three city blocks and parts of others, four acres of packed warehouses, factories and offices, a newspaper editorial summed up with the classic statement, "the fire department can only tackle a fire effectually pretty near the beginning or pretty near the end!"

Calamity on Bay Street

The street fire alarm box remained the principal harbinger of news of a fire long after telephones became common. This was especially so when fire gained a head-start in a closed building after hours and was first spotted by a passerby on the street as it finally burst through the windows. Many major cities in North America have a club of fire service fans who operate vans dispensing hot beverages and sandwiches to the firefighters at long fires. Some of these clubs are named after a fire alarm box, usually a famous one that sounded the alarm for the city's greatest fire. In Toronto it is the Box 12 Association. Box 12 was located at the hub of the business district, the corner of King and Bay streets. It is memorable because night watchman T.H. Johnson "pulled the hook" at 8:04 p.m. on April 19, 1904, an evening until then noteworthy only for a late-season storm of wind-driven snow, with a temperature of only 24 degrees Fahrenheit. Police Constable Bill Lannin also spotted smoke rising above E & S Currie's Men's Furnishings Store at 64 Wellington Street, and pulled Box 7, confirming the alarm. Toronto was about to have its greatest conflagration ever.

At the turn of the century, the downtown business district of Toronto was an important centre of commerce, finance, manufacturing and warehousing, all pretty much mixed together. It was second only to Montreal. The buildings had been built over a considerable period of time. Putting up a front of solidity in brick, cut and carved stone, they belied the vulnerability to fire that existed in their interiors.

Before modern multi-storey buildings, which have steel or concrete frames on which the walls are hung, the floors and roofs had always rested on the walls below them. When large rooms were needed, interior columns of wood or masonry had been used, but were more recently supplanted by cast-iron columns, capable of bearing great weight and resisting the heat of a fire better than steel. Although iron and steel supports and floors of masonry and concrete were making their appearance, the bulk of the three- to six-storey structures making up the core of Toronto had brick walls with wood-joisted interiors. Open stair and elevator shafts made chimney-like paths for fire, and wood-cased, plain glass windows exposed neighbours to one another's fires.

Automatic sprinklers, still the surest fire protection today, made their appearance some 20 years before, but were yet a novelty outside of a few large factory buildings. The sprinkler water curtain was an outdoor adaptation of this device and was used to protect the outside of a building. Both these forms of protection were to be significant in this conflagration.

Public fire protection was still weak. The water supply was dependent on an outgrown grid of mains in the streets, mainly from a 24-inch-diameter arterial feeder along Front Street. Many of the hydrants in this heavily congested district were located, however, on small 6-inch branch lines. This was not a system to handle the huge fuel load potential in the buildings.

The fire department of Toronto, Chief John Thompson had warned the Board of Control, had serious weaknesses and they were tempting fate by procrastinating over improvements. There were 15 firehalls, 16 hose companies, two aerial ladder trucks, a water tower, five city service ladder trucks, a salvage wagon, two chemical engines, 70 horses and 196 firefighters, but there was a weak link. There were only three steam fire engines and a deteriorating hydrant water supply. A number of much smaller cities had more engines than Toronto.

The running card called for a response to Box 12 of all three engines, seven hosewagons, an aerial truck, service ladder, water tower and salvage wagon, the heavy response typical of the day in such vulnerable districts. Also typical was the immediate response of the top chief.

In one of the most costly of Canadian business district fires, a long accumulation of human effort in Toronto was wasted in one tempestuous night in 1904.

The Currie Building, partly two and partly four storeys, was the typical brick and wood-joist building, a clothing factory and warehouse with open vertical shafts from basement to top floor. It occupied a space about 35 x 175 feet. Chief Thompson arrived to find the building heavily involved, the fire having raced up the open paths from the basement to spread into all floors. Now, on the breeze, it was coming out the windows and igniting the adjacent six-storey hat, cap and fur supply warehouse across a 12-foot alley.

Thompson positioned his forces front and rear, put in a second alarm for reinforcements, and went into the second building to try to stop the fire spread, but was trapped on the fifth floor by the fast-moving fire. The chief, with his crew, was forced to slide down a hoseline to safety. Thompson slipped and landed heavily, breaking an ankle.

It was a critical time to lose the chief, and there was some delay before Deputy Chief John Noble arrived on the scene. Two buildings, one of six storeys, were now blazing furiously in the wind accompanying the snow squalls. At about 8:30 the outside sprinkler water curtain on the Brock Building across Wellington Street, on the corner of Bay, was turned on to protect the windows from the fire, blown towards them by the 30 m.p.h. northwest wind. This successfully kept the fire away from this building, but the flames leapt across Wellington Street to neighbouring buildings on the west of it, bypassing the Brock property to push on through the block towards Front Street.

A general alarm, bringing out every available unit and firefighter in the city, was sounded at 8:51 p.m., as the fire had crossed Mincing Lane, was in the building facing Bay

Street and before long would be across Bay. The fast-growing conflagration was now burning in three separate city blocks. Chief Noble placed a heavy defensive line of 20 hose-streams here on Bay Street to hold the eastward spread, and the small size of the buildings to the north helped in this effort.

However, this was at the cost of southerly movement of the fire by ten o'clock to the rear of buildings facing on Front Street. At 11 o'clock the relentless inferno was driving east along Front Street. To the west, the Queens Hotel was saved, but the whole blockfront on the west side of Bay, from Yonge to The Esplanade, was reported to have become totally involved in flames in 45 minutes. In this case it was ignition by intense radiant heat, not flying shingles, that demoralized the firefighters by leapfrogging past their defenses.

Thick crowds gathered in spite of the weather, a mix of owners and employees wanting to rescue their goods, and sightseers. Outside help was called. Soldiers came to dynamite buildings, but this was thought better of and they stayed to help in conventional firefighting and crowd control. Suburban firemen came in, as did two hose-and-engine companies with 30 men by train from Buffalo, and 12 men with 2,000 feet of hose were sent from Hamilton. There was also a contingent from London.

Firefighters were hampered by the inability to develop enough powerful streams, the radiant heat, wind and festoons of overhead electric wires made it difficult to get water on the fires. At midnight the wind shifted to the north. The fire crossed Front Street to burn a number of buildings on its south side, and crossed Bay Street here also, the intersection becoming a vortex of heat the firefighters could not approach. The flames then commenced to eat their way north back up Bay Street to add more buildings to the toll.

A determined effort was made to keep the burning flour mill and manufacturing plants on Front Street from spreading the fire east to Yonge Street, and this was successful. The firewall protected Minerva Block on the north, and a windowless masonry wall on the south gave the exhausted Toronto firefighters the help they needed. The ornate Customs House, the next victim in line, was thus saved. Foreman Frank Smith led the desperate battle to contain the conflagration on this southerly front.

The large Kilgour paper box plant on Wellington was protected by automatic sprinklers inside and water curtains outside. The protection saved this key property from the advancing fire, though a three-storey furniture warehouse adjoining it was burned to the ground. The outside water curtain sprinklers weakened as the city main pressure dropped and fire got in, but the inside sprinklers, served by rooftop tanks, stopped it.

By 3 a.m. the fire had travelled north on Bay Street back to the Brock building, earlier saved by its water curtain on the Wellington Street side. Now, ironically, the fire entered by the unprotected windows fronting on Bay. Inside sprinklers operated, but the rooftop gravity tanks gave out and the structure finally succumbed. To the north of this, on the east side of Bay Street, the sprinklered *Telegram* newspaper building and its employees efforts helped stop the fire in its progress north.

Flying embers rained down on The Esplanade, where the Gardiner Expressway is now, and on the harbour shipping along it, which fled, but no serious fires started in this area, occupied largely by railway trackage. The Great Fire of Toronto was stopped and under control at about 5 a.m. Flare-ups and wetting down of hot ruins would go on for another two weeks, as fire burned under the rubble. Some safes and vaults protected their contents, others failed. All had to be left until their contents cooled down, so they would not burst into flames on exposure to air. Businessmen bit their nails and waited in the interim.

In this display of the might of fire, unleashed there were 98 buildings destroyed, all brick- or stone-walled with wood interiors and averaging four storeys, the area covering about 20 acres. The loss was estimated at $10,500,000, largely insured, unlike most previous conflagrations; some $8,400,000 in claims were paid out. There were no fatalities, but many injuries among the firefighters. They were reported to have worked heroically against great odds with limited resources until they dropped or were blown over by the wind.

Typical resources available to a modest sized city in the day of horsedrawn fire apparatus. A turnout of the Hamilton, Ontario, Fire Department.

Like the major conflagrations in various other central business districts of the period — Baltimore, Rochester, San Francisco and Montreal — this fire showed that their big masonry-walled but fuel-packed buildings would radiate sufficient heat through their windows to leap most streets and enter the windows of the buildings opposite them. The manufacturers of wired glass made a great deal of this and prospered, as the heavy chicken-wire in a glass sandwich set in a metal sash withstood fires yet admitted light as a window should. Dependence on sheet iron shutters that had to be opened and closed each day became passe, while sprinkler water curtains gained wide acceptance.

The theory of the firestorm in World War II coinage, is of a vertical column of hot gases rising from a large-area fire that results in strong winds rushing inward to the base of the column, like the draft into a fireplace. However, there are indications great fires like Toronto's create or increase horizontal surface winds as well. Certainly forest fires evidence this phenomenon.

In addition to finally listening to the fire chief's pleas for a stronger fire department with more men, equipment and especially steam fire engines of large capacity, a special system of high-pressure water mains was installed in Toronto, utilizing extra-large pipes, hydrants and big pumps rated at 300 pounds pressure, duplicating and surpassing the regular system. It was put in service in 1909, covering the entire commercial and industrial core of the city, some 300 acres. It was never utilized to anywhere near its full capacity before being converted to regular domestic supply system sources and pressures in 1980.

The last time the system was used to help control a severe fire was for the Eaton's Fire in May 1979. The old Eaton department store on Yonge Street and a group of the company's former warehouses were under demolition when a wind-swept fire raced through some of the old structures, spreading to other buildings and the Old City Hall. The old Eaton buildings had been well protected by automatic sprinkler systems for many years, but, like all buildings under demolition, were a serious hazard. The largest fire in Toronto since that of 1904 was the result, the central financial district being showered with burning embers. Conditions had changed, however. The high-pressure system was not utilized to anything like its capacity, powerful motor pumpers supplying most of the large-calibre streams of water needed and the modern buildings resisting ignition.

(XVIII) Facing the Challenges

Joseph Tremblay, Superchief

Joseph Tremblay was a big, purposeful man. Tailoring was evidently not stimulating enough for him, so he gave it up to join the St. Cunegonde Police-Fire Department on the lower west boundary of Montreal, soon becoming its chief. When that suburb was annexed in 1905, Tremblay was made a district chief in the Montreal Fire Department, but shortly after was made first deputy chief in the department. His meteoric rise continued when he succeeded the notable Zepherin Benoit as chief of department in January of 1909.

Tremblay was a strong-minded man, and in these years of great change he had a stormy career, particularly in his relations with the somewhat despotic Mayor Mederic Martin. During his tenure, the chief handled a great expansion of the fire department. There were nine suburbs annexed in 1912 alone, the number of fire stations doubling in a ten-year period. He was not shy about recommendations, asking for 12 new fire stations in one year.

The first fire prevention division was founded by Tremblay in 1914 with an astounding 37 inspectors, but he promptly lost 25 of them as replacements for firefighters who went off to war. Tremblay saw a training school established under the storied Captain Doolan, a colourful hero, at No. 4 Station on Chaboillez Square. He brought in the motor car and adopted motor apparatus with enthusiasm.

The First World War was a time of bickering and near bankruptcy at city hall. Consultants reported that the fire department was the only one being run efficiently. The police were demoralized. This resulted in the appointment, in 1918, of Tremblay to the new post of director of public safety, ruling both departments. He chose as his deputy the very able Arthur Mann, a third-generation firefighter. Joseph Chevalier, the chief engineer, became fire chief under Tremblay.

However, it was not to last. In this very year, on December 10, both firemen and policemen went out on strike. For the firemen, as in a number of cities at the end of the war, there was a demand for union recognition and two shifts. But the police were gunning for Tremblay. On the 12th there were serious riots, robberies and holdups. The strike was settled on the 14th with the intervention of the premier, the strikers winning their principal claims, among which were the immediate and premature retirements of Tremblay and Mann. The superchief was subsequently successful in the fire apparatus business.

On the urging of Tremblay, the city had begun to award fire department medals for heroism in June 1914. Most awards on this first occasion were for service at the *Herald* disaster of 1910. Among the heroes were Ferron, Doolan and Mooney. Arthur Ferron was credited with 15 rescues and received three medals, one of them from the Humane Society for wading up to his neck through flood waters to rescue a family. He was also credited with releasing many trapped children in the Laurier Palace Theatre panic by breaching a wall.

The famous Chief Doolan, first full-time instructor in the department, dove underwater to connect hose to hydrants on McGill Street during one of the Spring floods Montreal suffered in those days. On one occasion he is said to have rescued a woman over a ladder on his shoulders while carrying her baby in his teeth. Lieutenant Joe Mooney was an athletic star with the Montreal team at the International Firemen's Congress in London some time before. he made several spectacular fire rescues, such as when he brought out Chief Benoit and two other firefighters, having gotten to them through his acrobatic skill.

The Mayor of Montreal presents a fire service medal to mechanic Raymond Pare, eventually to become fire department director. Director of Public Safety Joseph Tremblay reads the citation at left.

The collapse and burning of the Herald building in Montreal in 1913 was one of the more grisly of Canadian fire disasters. Chief Tremblay is pointing where the firemen should dig for bodies.

The Herald Disaster

One of the oddest and most tragic fires in Montreal was a bizarre occurrence that destroyed the *Herald* newspaper building on June 13, 1910. It was claimed that water caused the fire that resulted in 32 deaths and many injuries, while a hundred more were saved by the fire department.

About 300 people were working in the *Herald* building on the south side of Victoria Square that morning. Suddenly, without any warning, the elevated water tank for the automatic sprinkler system had its supports give way and came crashing down. The 60 tons of tank and water smashed through the roof and three floors to the ground-floor level of the busy newspaper and job printing plant, joined along the way by heavy linotypes and printing presses.

Fire broke out immediately, facing the fire department with the dual problem of firefighting and rescue of many trapped employees. The stairs were wrecked and the sprinkler piping broken. Many of 68 girls working on the fifth floor were brought down ladders, while numerous individuals had to be pried out of the wreckage. Courageous firefighters worked feverishly in the smoke and confusion, seeking out the moaning victims. Hospitals became jammed. Many were the stories of heroism, tragedy and escape. Some were killed in the collapse, some burned to death. One old linotype operator had barely escaped in the burning of the previous *Herald* building many years before. Another operator fell with his machine four floors and walked out the door unhurt.

It took a week to recover the bodies of all the victims from the burned and wrecked shambles, though from the street the building seemed little damaged. The specific cause of the collapse was not formally decided, but the building was evidently heavily loaded with printing machinery on the upper floors.

Calgary's Cappy Smart

Born in 1865, son of a Scottish sea captain, James "Cappy" Smart immigrated by cattleboat to Canada, arriving in Calgary in 1883. Smart was a volunteer fireman in Calgary in 1885 and was elected chief in 1898 on the resignation of Chief McClelland. In his early years as chief he was also the town's only undertaker.

Before becoming chief, Cappy served as fire prevention inspector; as chief, he continued to promote it, becoming a pioneer in the regular inspection of homes and commercial establishments by the regular firefighting force. In 1903 there were seven full-

A colourful fire chief in Calgary's history was James "Cappy" Smart, who converted the fire department to motor apparatus very early. Here he poses in his touring car with other early motor trucks.

time paid firefighters, including the chief and a volunteer force of 30. By 1909 the volunteers were gone and there were 30 career firefighters manning the two stations.

Calgary was one of those rapidly growing western cities, by 1912 requiring seven fire stations. Chief Smart ordered inspections of all buildings by the firefighters in 1916, the beginnings of serious fire prevention, though it would be many years before a fire prevention bureau with full-time specialists would be formally established. For a time, Cappy's daughter Minnie was his secretary, one of the first female employees in a Canadian fire department.

Smart also propelled Calgary Fire Department into the motor fire apparatus era early and in a hurry. First was his car in 1910, followed in the next few years by a collection of pioneer motor apparatus supplied by R.S. Bickle. The chief's open touring car almost finished his career when he ran broadside into a streetcar in 1912. It took two years for him to recover, and he thus missed the week-long Burns Packing plant fire in 1913, the longest and toughest the department ever fought. A Waterous steamer pumped for the whole week in -30 degree Fahrenheit weather.

Calgary made runs to outlying towns, by train to High River with horse-drawn apparatus, and later by road with open motor apparatus without windshields. In 1921 the department sent a pumper and crew by C.P.R. extra to fight the fire that destroyed the old Banff Springs Hotel. This time they travelled in style, in the business car of D.C. Coleman, district superintendent of the line, with the services of the superintendent's personal chef. It was, however, a wild ride at maximum speed, and the six-hour battle did not save the hotel.

Cappy Smart was the sort of leader who was tough but had respect for his men. He believed a firefighter had to eat smoke and like it. Smart was very much involved in whatever was going on in Calgary and was for many years the official starter of the annual Stampede parade. One of those characters of legend, he fought fires hard and had many scars to show for it. Of course some of these may have resulted from such escapades as diving through a window to escape being caught by the police at an illegal cockfight.

Chief Smart guided his department through a period of rapid growth, an old-time chief who always led his men into the fray personally. It did not occur to his breed that they might be more useful standing on the street directing the whole operation. With them the job was to lead, not direct.

Cappy Smart ruled the Calgary Department as active chief until 1933, when he retired to take the title Honorary Fire Marshal, which he held until 1935, allowing him to claim 50 years of service to the fire department and earn the King's Medal for fire service. He died at 74, four years later. Cappy was succeeded by Alex Carr, a veteran of 31 years of service.

A New Century for Toronto

Toronto strengthened its fire department substantially in the years following the disastrous conflagration of 1904. In addition to this calamity, the firefighters had had their own personal tragic reminder of the perils of their profession in 1902. An explosion occurred during a fire in the McIntosh Building at Front and Esplanade streets on July 10. Three members of No. 5 Hose Company were among five men killed. Others were overcome in vain attempts to save them.

The high-pressure water main system was first used on a big fire in a paper box plant on Scott Street early in 1909, a general alarm fire. Four big streams were utilized under the full 300 pounds pressure available in this system. It was used again when the west wing of the Parliament Building, including the library, was gutted. The premier himself turned in the alarm and, with some of the members, helped rescue valuable pictures.

The big, impressive, six-door station that remained headquarters for over 60 years was opened that year on Adelaide Street, west of York. By 1912 there were 27 firehalls, housing 338 men, 120 horses and 55 horse-drawn wagons, trucks and engines, and there were 392

The Toronto Fire Department recognized the climbing toll of its firefighters lives in the late 19th century, producing this montage including Chief Richard Ardagh and ten firefighters.

fire alarm boxes on the streets for fast sounding of alarms by the public. At this point the first piece of motor fire apparatus was delivered, a Seagrave air-cooled hose and chemical truck. Rapid motorization of the department with more of these units, LaFrance pumpers and ladder trucks of both makes ensued.

Chief Thompson had been popular and championed the welfare of his men, but could not come to accept the formation of a union apparently. This would seem to have led to his retirement in 1915, the more sympathetic William Smith replacing him. The union was recognized and the two-platoon shift system adopted in 1918, raising the firefighting personnel from 342 to 522 men. Chief Smith retired soon, in 1919, having served as a fireman for over 50 years. He was succeeded by William Russell, who stayed for 11 years.

Always a protection problem for the Toronto Fire Department were the Toronto Islands, the sprawling sandbars with lagoons and channels among them, a recreation area and the shelter that created the harbour. For many years it had the typical crowded, flimsy amusement-park-style buildings as well as many separate areas of narrow residential streets of lightly built cottages. The islands were a ferry ride away from the rest of the city.

For this reason, even with one fire station on the islands, retention under contract of a tug with firefighting capabilities was a practice in this era. It was available for fires on the islands, in the channels and the main harbour, as well. The contracted boats were replaced by a fire department-owned fireboat, rated at 1,200 gallons per minute and designed to navigate the small channels, in 1923. It was named *Charles Reed* and operated from the island fire station.

The built up area of Hanlan's Point on the islands was visited by a conflagration in 1909, for which steam fire engines responded from the city by scow. With inferior construction and delay in getting reinforcements to the scene, 17 buildings were destroyed.

Just before World War I the major cities began to try out motor fire trucks, in this case a Seagrave, aircooled hose and chemical wagon known as a "buckboard".

The First World War

Manning problems were caused by the outbreak of the First World War as ablebodied career and volunteer firefighters alike went off in droves to fight for King and Country. In the cities, substitutes were hired for the duration, the soldier firefighters usually being promised their jobs back on their return. It would appear, though, that the adoption of the two-platoon system of manning in the bigger cities must have absorbed many of these substitutes in 1918 and 1919.

The rush of men to sign up in Montreal cost Chief Tremblay most of his new fire prevention bureau. In only a few cases are there reports of significant delays in the acquisition of fire apparatus, considerable numbers of motor trucks being placed in service to replace horses. This is possibly due in considerable part to the fact that the components for many fire trucks came from the United States, which did not go to war until 1917.

There was a tendency to cry sabotage and see enemy spies under every bush following any particularly large fire, but no such enemy activity was authenticated. Two particularly noteworthy major fires occurred in Canada during this war, the burning of the Parliament buildings in Ottawa and the fires that ensued following the Halifax explosion.

The Burning of Parliament — Again

Canada's united province had had calamitous fires in its Parliament buildings in Toronto, Quebec and Montreal. Now the Confederation of the Dominion of Canada was to repeat the performance. The Parliament Building complex built in 1859-60 in Ottawa, the new capital of the Dominion, marked the final settling down of the Canadian seat of government in a permanent location. It was one of the most impressive government legislative blocks in North America, a $6,000,000 solid-looking 470-foot-long group of sandstone buildings with a 220-foot central tower and numerous smaller towers in a busy Gothic style. A circular library structure was added at the rear at a later date. The buildings stood where the present Parliament Building does, on the high bluff overlooking the Ottawa River and on a rise above downtown Ottawa.

The 1860 building was a refreshing change from the Roman domes of so many other capital buildings, but it was built with a fine, total disregard for structural fire integrity. Aside from the solid stone walls, the interior consisted of wood, a central building and two wings interconnected by a labyrinth of combustible passageways and vertical openings, trimmed in pine, without any form of cutoff. Inside, the impressive pile was a bomb.

The Ottawa fire chief had urged better fire protection without any action being taken to remedy the obvious vulnerability of the centre of government. There was an alarm system, portable extinguishers, and a standpipe and hose system, but these left too much in the hands of amateur firefighters. A detector system was in use, with a relay to the city fire department.

The fire began at about 8:55 p.m. on February 3, 1916, in a newspaper reading room on the first floor. This room was filled with racks of newspapers, books and piled papers. The furniture and fittings were in oiled and varnished white pine. A member of parliament, who was in the room, suddenly felt a blast of hot air and smelled smoke. He looked up to see smoke and flame billowing from one of the reading tables. He called a police officer, who attempted to put the fire out with a hand extinguisher. This failing, a standpipe hose was brought into play, but the flames spread undeterred across the papers and combustible woodwork that had been drying for over 40 years.

The fire rolled with great power along the flammable trim of the corridors of the House of Commons and upward into the large roof space over the House. There were no fire doors or stops in the complex to serve as fire barriers, except at the entrance to the semidetached library.

The expression "built to burn" described the original Dominion of Canada Parliament building of 1860, despite its solid stone walls. Burn it did, in 1916, the flames racing through extensive interior concealed spaces. - PAC-PA9234

The House was in session and the members and visitors were obliged to make a rush for the exits as smoke billowed into the chamber. Some in the gallery could not use the stairs and were rescued by ladders. Two women in the gallery were not taken out for half an hour and died later. Accounts vary, but it has also been reported that a member and a law clerk were trapped and burned to death, and that three employees were killed when the central tower fell as they were fighting the fire at 1:30 a.m.

The Ottawa Fire Department received an alarm from the automatic alarm system at 8:57 p.m., and the first company was on the scene in two minutes, but the flames were already through the roof. The fire had flashed through the huge roof spaces and dropped through the glass ceilings of the house of Commons and the Senate chamber.

Additional alarms brought all available fire companies, which used all the hydrants on the hill, mustering 20 streams of water. They had little chance of staying the force now unleashed, and the fire made its way through most of the massive complex, not being brought under control until about 4 a.m. Only the library was saved, remaining to form a distinctive part of the present Parliament Building, the old one being demolished and replaced entirely. The loss was set at $3,000,000, uninsured.

The Halifax Explosion

On the clear crisp morning of December 6, 1917, the Belgian relief ship *Imo* was putting to sea, sailing from the anchorage of Bedford Basin, down through the narrows between North End Halifax and the Dartmouth shore. Ahead was the broad reach of Halifax harbour and the Atlantic. Heading inward through the narrows was the French freighter *Mont Blanc*. Unaccountably, something went wrong with their adherence to the rules of the road. Action to avoid a collision became imperative and both ships took confused, evasive measures that failed. They collided, *Imo* slicing a big gash into the forward starboard side of *Mont Blanc* in No. 1 hold. Still, it would have been a relatively minor "marine casualty" but for the fact that the *Imo*'s victim was an ammunition ship.

There was a war on, and Captain Lemedec did not choose to fly the usual red explosives flag, advertising to any U-boat what his cargo was. Known to almost no one ashore was the fact that *Mont Blanc* was stuffed with 2,600 tons of explosives, including 225 tons of TNT, 2,300 tons of picric acid, 61 tons of other acids, and a deck cargo consisting of drums of benzol.

On impact the drums of benzol were thrown about, burst open and the gasoline-like liquid sprayed down through the gash into the cargo of shell-explosive Lyddite and other nasty things occupying the 'tweendecks' below. Sparks from the grinding steel of the two hulls ignited the benzol or the picric acid, causing a billow of black smoke and red flame, punctuated by minor explosions.

Captain Lemedec knew with mortal fear what he was facing. The whole foredeck was ablaze; he could not anchor. He could not scuttle the vessel, as the seacocks were seized up. The captain and his crew took to the boats and rowed furiously for the Dartmouth shore, leaving the floating superbomb to drift, its fuse well alight.

The story of the dramatic disaster that followed has been told many times. The fire aspect, however, has been treated vaguely and the record of it is incomplete. The high drama was, after all, in the enormous split-second cataclysm said to be the greatest man-made explosion before the atomic bomb.

Knowledgeable observers of the burning *Mont Blanc* soon knew there was a disaster in the making, as she slowly drifted towards the busy docks of Halifax and *Imo* made haste from the scene towards Dartmouth shore – the flames turned from red to blue as the oxidizers intensified the fire. Officers in H.M. Cruiser *Highflyer*, anchored nearby, saw it and reacted without hesitation. The steam pinnace was away, so a whaler and boarding party set forth and boarded the burning ship. The small harbour tug *Hilford* also went alongside, and an attempt to tow the blazing *Mont Blanc* away from the docks was begun.

Employees of the dockyards, nearby industrial plants, and citizens on the streets and in their homes watched the spectacle in the harbour from roofs, windows and the fine vantage points on the slopes at Fort Needham and the residential district of Richmond. They were all ignorant of the threat to which they were exposed.

Just before 9 a.m. a grocer, Constant Upham, thought to telephone the fire department. About the same time, someone pulled fire alarm box 83 at Pier 6. *Mont Blanc* had now brushed this wooden pier, setting fire to it.

The Halifax Fire Department responded to the call, dispatching the horse-drawn hosewagon from nearby Isleville Station on Gottingen Street, backed by the first and only motor apparatus in the department. This was the first LaFrance pumper in Canada, the triple combination named Patricia, running out of the big downtown Bedford Row Station with a six-man crew. Fire Chief Edward Condon and his deputy, William Brunt, followed Patricia in the chief's touring car.

The sledgehammer blast threw the wooden houses, schools and churches of the Richmond district off their foundations, flattening, splintering, tearing them apart. Fires broke out in some cases as stoves upset, lamps fell and wires shorted. Spectators were thrown off roofs, and shattered glass was hurled into the faces of the hundreds who had been watching through their windows. In the vicinity of the dockyard industrial area and old lower Richmond streets, buildings collapsed, burying their occupants. The stunned city was, in an instant, confronted with blazing docks, sugar refinery, cotton mill, foundry, naval dockyard buildings, and an unknown number of homes, many with occupants trapped in them. Many of these imprisoned people were doomed to death by fire, if they did not die of exposure.

It was not only a Halifax disaster. The north end of the Town of Dartmouth, across the harbour, also suffered severe damage. The Episcopal church and the skating rink collapsed. A considerable number of houses were wrecked and a small Indian reservation virtually wiped out. Prominent commercial establishments suffered severe damage. The Halifax Brewery and its staff were obliterated, being directly opposite the explosion point.

The fire apparatus had apparently just arrived on the scene at Pier 6 when the inevitable happened. Presumably, the trinitrotoluene, 450,000 pounds of it, detonated, taking along with it 2,000 tons of other explosive materials. The cataclysmic blast, according to many witnesses, came as two shocks, the first a silent concussion, then a great, grinding, roaring explosion. There was a blinding flash and a mushroom cloud of smoke, first black, then turning grey and white, that pushed an estimated three miles into the heavens.

Mont Blanc was obliterated, small parts of her landing up to several miles away. The tug *Hilford*'s engines were found far ashore, nearby ships had their superstructures flattened. The big pier sheds and industrial plants were shattered and in flames; railway boxcars were thrown about like toys.

Chief Condon's car somersaulted backward and lay upside down, both chiefs dead. The Patricia was stripped of her sheet metal work and battered out of action, all of her crew killed except one. Billy Wells, the operator, flew off her with part of the steering wheel in his hands. One of the Isleville horses' collars is said to have been found two miles away. A tidal wave of harbour water then poured over the waterfront wreckage, adding to the damage and the death toll. Billy Wells was carried along in it, almost drowning, but came to amongst the wreckage.

"Patricia" was the first and only motor pumper in Halifax. Arriving at a reported fire on the Halifax waterfront, she was caught in the great explosion of 1916 and most of her crew killed.

Detailed fire department records are lacking. Accounts state that a second alarm brought out the remainder of the Halifax Fire Department's forces. A confusion of alarms is probable. The North End alarm box circuits would have been smashed with the poles that carried them. In any event, it would not have been long before all companies and off-duty men turned out. The naval vessels in port, and the regular army and militia units quickly turned to, as did parties from two American warships in port. Another American ship was utilized as a hospital to help the shore hospitals with the massive medical load. The major jobs demanding attention were rescue of trapped people, some of them in burning buildings, transporting and treating the injured, and firefighting.

The fire department had lost its two commanding officers. Station officers seem to have worked individually, going from one fire to the next, trying to do as much as they could. City Controller Hines took command on the key southern side of the devastated area, assuming for some hours that the chief was working on the north side.

Most accounts suggest a large-area conflagration ensued following the explosion, and eyewitnesses indicated this by stating they could see nothing but smoke and fires in the Richmond area. However, such little concrete information as is available and examination of a number of photographs of the devastated area suggest that there were many individual fires in damaged and wrecked structures, but the spread from building to building was relatively limited. The houses in Richmond were fairly widely detached. It was in the more congested lower levels around Barrington, Union, Gottingen and Russell streets that groups of houses burned together. Those shattered large industrial buildings that were relatively intact along the North End waterfront were left to burn out after rescue attempts.

The rock promontory of Citadel Hill, in the middle of the city, deflected the blast from much of its area, toward the South End, but there was some fear of a conflagration sweeping into the denser parts of the city. There was no wind and the most threatening fire, on Gottingen Street near the Wellington Barracks, was handled by the firemen and some naval parties — a grocery store had caught fire and the neighbourhood was threatened.

A rare picture of firefighting operations in Halifax following the 1916 explosion. The Shand Mason steamer is working in the area of Gottingen St.

The Halifax Fire Department forces included the one wrecked motor pumper, five active and two reserve steam fire engines, two chemical engines, four ladder trucks and eight hosewagons, all horse-drawn. They were manned by 32 career firefighters and 120 call men, two of whom were assistant chiefs. They went to work as professionals to fight fires and rescue trapped and injured people, tackling whatever jobs faced them, knowing it was very little related to the whole disaster scene. A key point for the firefighters was in the area of the Wellington Barracks at Gottingen and Russell streets, the limit of heavy blast destruction and a frame-built area where the fires might have spread into downtown.

The measure of the catastrophe is in the bare statistics. Officially there were 1,960 dead — substantially more by some accounts — 9,000 injured, 6,000 homeless, 1,600 buildings destroyed and 12,000 damaged. The dollar loss was about $35,000,000.

A panicky flight of citizens to the south end of the peninsula occurred when rumours of the imminent explosion of the military ammunition magazine in the city were spread, but the firemen stood their ground in the area, continuing their job of firefighting. The rumour turned out to be greatly exaggerated, although all combustibles were cleared from the vicinity by troops and sailors in case the flames should reach it.

Statements were made by witnesses that a massive fire seemed to be raging from the docks, through the Richmond district to the top of the hill behind, and that the North End was "a vast burning scrapyard." St. Marks Church blazed intensely and thick smoke was everywhere, but the reports by those at a distance must be taken as somewhat less than a precise representation of the actual condition. The fires, in any case, reached a peak between 11 a.m. and noon.

Long distance assistance at major fires in the era of horses meant special trains. This one was the response of a steamer and hosewagon from Yarmouth to Annapolis Royal, N.S.

Royal Navy sailors and marines were active in the rescue work, hurriedly digging injured and trapped victims from wrecked and sometimes burning houses, having to abandon some to death by fire and concentrate on the ones they could get out without delay. Among the many unidentified dead only 63 were described as charred, these recovered mostly from Barrington and Albert streets, and the drydock, railway and foundry ruins.

The news spread rapidly and relief trains carrying doctors, nurses, supplies and firefighters were underway within hours. A train from New Glasgow with a steamer, 3,000 feet of hose and 40 firefighters was made up in an hour. It stopped at Stellarton to pick up more firemen, and ran down to Halifax in the early afternoon. Similar trains came from Sydney, Amherst and Kentville. All arrived by early evening.

However, the fire situation, as well as the urgent rescue work, was under control by about 4 p.m. Sailors were relieved of their firefighting activity at 2:30 p.m., when advised by a fire department officer that the situation was in hand. The fire crisis, however serious it may have been, was over. New Glasgow and Truro firemen overhauled ruins and patrolled the streets of the North End during the night while the Halifax firemen rested.

All fires were out before morning. By then, a driving blizzard was getting underway, adding to the misery of both the searchers of the ruins and the thousands whose houses were damaged or without window glass. The two chiefs and six firefighters lost their lives in the discharge of their duties.

Permanent physical and psychological scars remained with Haligonians. In World War II, any loud bang stirred memories and fears, as another generation of munitions-laden ships sailed by at the foot of their streets. A junior officer of the Halifax Fire Department at the time of the explosion, Lieutenant John W. Churchill, was placed in command to succeed Condon, remaining as chief through the twenties, thirties, and the dangerous and trying episodes of World War II — 28 years in office. Because there were no fire chiefs in the Halifax department until 1894, he was only the fifth chief when he died in office in 1945.

Firefighters' Unions and Two Platoons

The long-suffering career firefighter had always accepted the continuous duty system. It was part of the job then. When most other working men put in a hard ten hours work, six days a week, the fireman had many hours to put in just waiting. Looking after their beloved horses, taking little practice, reading, playing cards and sleeping, all was subject to interruption at the clang of the bells. For about $70 a month, the firefighter typically was on duty in the station 21 hours a day, taking an hour to go home to each meal on a staggered basis. During these hours the remaining crew ran to fires short-handed, remaining so until the others arrived from their homes. Each man would have off something on the order of a half-day and a night each week, if he were lucky and no one was off sick. In some departments meal hours were extended to allow church attendance on Sunday.

The First World War changed this, as it shook up the world in general. Firefighters returned from the war expecting a better deal. It was also a time when unions, generally, were finally gaining considerable ground in the reduction of working hours and improvement in conditions in the workplace.

The question of work hours for firemen came to a head in a number of major cities when the servicemen returned. Civic officials resisted the increased cost and, in 1918, firefighters in Montreal, Winnipeg and Vancouver went on strike to reinforce their demands for two separate shifts that would alternately man the fire stations. This would give them an 84-hour duty week and was referred to as the two-platoon system. In Winnipeg the firefighters also joined the violent general strike that paralyzed the city for a lengthy period in 1919.

In the case of Montreal, it was a combined strike of police and firefighters, starting on December 10, 1918, and lasting four days. The secondary demand in Montreal was unusual,

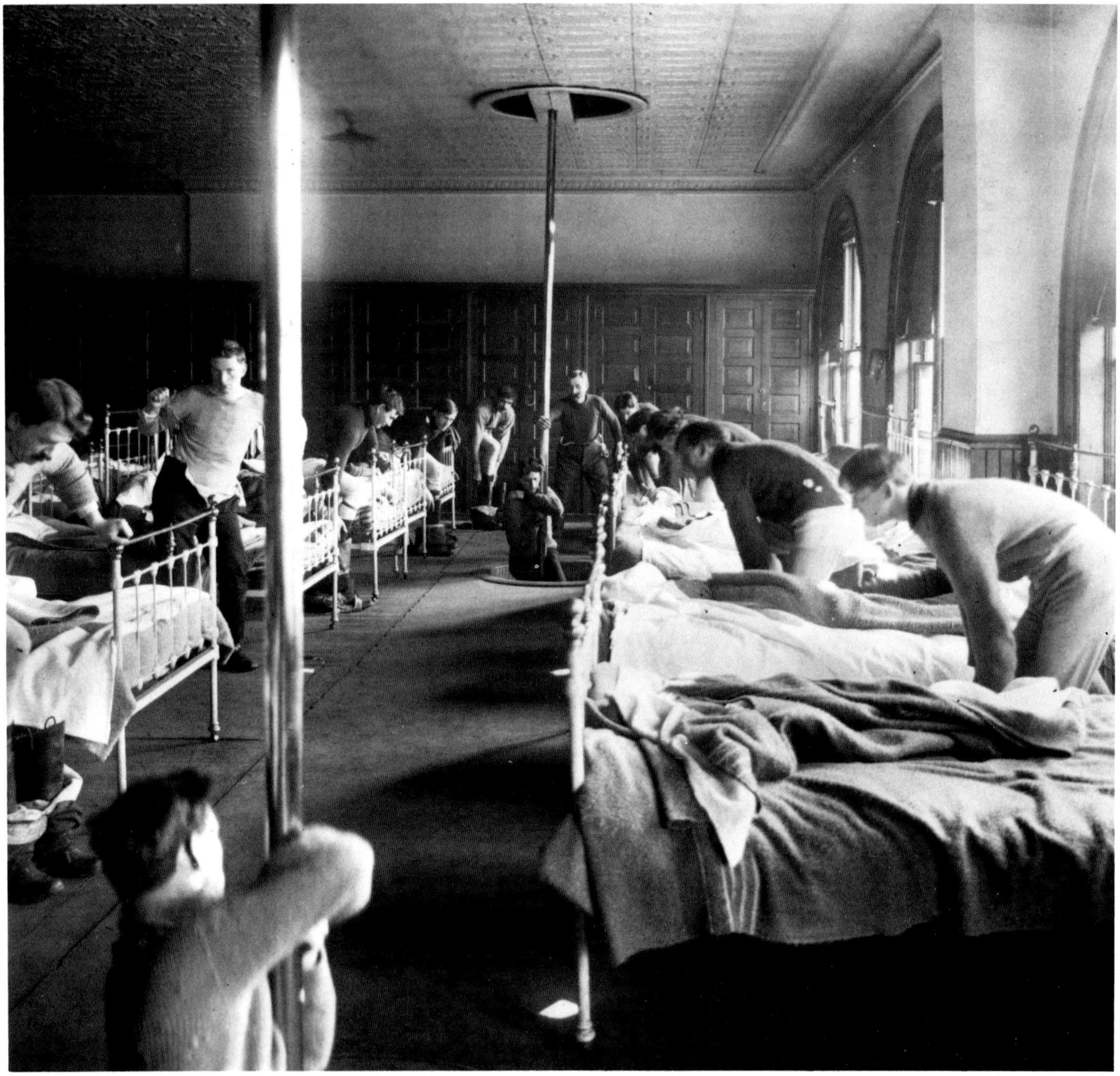

Dating from the days when they lived continuously in their fire stations, firefighters have slept during night shifts, turning out rapidly on alarms in a "night hitch" of clothing. - PAC-PA29716

the dismissal of Chief Joseph Tremblay, recently promoted to director of public safety over both departments.

Tremblay was breaking up the politics-ridden sinecures that made the police ineffectual. He was too tough for the police and made enemies. They decided he must go. To back their propaganda against Tremblay, they accused him of being a Mason, then a damning charge in Montreal. He was not a Mason, but the city capitulated, granting the two-platoon system to the firemen and firing one of the best chiefs they ever had, leaving the unprepared former chief mechanic, Joseph Chevalier, to head the once again separated fire department. The firemen also won union recognition and more money.

In most of these strikes, officers and volunteering citizens tried to provide fire protection for their cities, causing some bad feeling. However, the citizens demonstrated that they could, if they set their minds to it, avoid having fires at all, and few serious ones occurred.

Firemen had long had their own benevolent societies, through which they helped one another and their brothers' widows and orphans. Now they began to fight for greater contributions from their employers.

Before the two platoon system came in the firefighters put in very long hours in the fire stations. These Vancouver crews are in typical surroundings for 1910.
- PAC-PA31266

The sinister black mass of the 1916 forest fire storm races down on the rough bushtown of South Porcupine in the Northern Ontario forest. - PAC-C29808

6

A DIVERSITY OF FIRE PROBLEMS (1920-1945)
(XIX) A Great Loss of Life

Firestorms in the Bush Towns

New towns carved out of the frontier forests have always had the special risk of forest fires running into their communities. This has applied to towns centred upon the lumber industry and to mining towns, notably in Northern Ontario, where the realization of these fears came repeatedly. Existing in the forests, the towns were naturally built almost entirely of wood. Enormous losses of life, property and timber have resulted from a few large-area forest fires.

Descriptions of the Great Miramichi Fire in New Brunswick, and the blaze that so quickly dissolved the fledgling City of Vancouver, have given a hint of the fury possible in a forest fire under the right conditions. The Great Fire of Chicago in 1871, the one that made Mrs. O'Leary's cow famous and cost 300 lives, received worldwide attention and was very destructive indeed. Even in American newspapers it eclipsed almost entirely the more cataclysmic Peshtigo Fire, as it is called, in Wisconsin. This fire cost probably more than 1,000 lives, destroying a number of frontier lumbering towns, notably Peshtigo, which was wiped out in a matter of minutes by a firestorm travelling at over 100 miles per hour. The leading edge of hot gases was estimated at perhaps 2,000 degrees Fahrenheit.

This huge forest fire, and possibly others spawned from it, raged through three American states and Northern Ontario, with the extent of the calamity not known to the outside world for a week. Some 2,000 square miles of unsettled timberland in northwestern Ontario were burned over. The Peshtigo fire and the series of fires that burned out a number of towns and settlements in the Saguenay district of Northern Quebec in June 1870, reiterated the fact that fires could travel with incredible speed and intensity during a sunbaked summer in virgin forest, or where land clearing has caused accumulations of dried-out slash. Unfortunately, the lessons did not stop the carelessness of campers, lumbermen and railroaders.

There was little or no control by government over the hazards created by settlers and railway or road builders in their hurry to bring civilization and exploitation to the endless forests of North America. It would not be until well into the twentieth century that any real action was taken. Though villages learned to clear the brush back from the edges of town, it was seldom enough to stop a large-scale forest fire.

About 80 lives were lost and 3,000 homeless refugees sent fleeing to Cranbrook when forest fires raged through the Elk River district of British Columbia in August 1908. They completely destroyed the towns of Fernie and Michel, as well as causing considerable damage at Hosmer and Coal Creek. It was said that, as the fire raced up the Elk Valley, flaming tree branches five and six inches in diameter were carried over half a mile on the wind. At Fernie, radiant heat ignited the frame buildings at a considerable distance. Property loss was estimated at $4,000,000.

Northern Ontario has experienced a number of very large forest fires that destroyed numerous sizeable towns, some being caught twice. Between July 9 and 11, 1911, the towns of Cochrane, South Porcupine and Pottsville, as well as lesser frontier settlements, were burned out. The area mining and lumber camps were destroyed and hundreds of men, women and children forced to take refuge in lakes and streams. Some of them drowned, and others in the outlying bush perished without leaving a record. It is estimated that several hundred people died in this fire. Not only were the buildings of wood, but every building had its firewood supply piled alongside, seasoned and very flammable. As in Peshtigo, the townsfolk ignored the alarms raised by refugees from the fire, who rushed into town ahead of the advancing firestorm.

Even larger forest fires swept through the same north country in the exceptionally dry summer season of 1916. The coniferous forests were like tinder in late July, when a number of local fires combined to create a tidal wave of fire that roared down on numerous towns and villages. About 243 persons lost their lives, while those who could dive into a lake or muddy pond sometimes survived.

The fire bore down on the rebuilt Cochrane about 10 a.m. on the 29th, but at first the fire department and citizens held it at bay. Another, separate fire then started in the muskeg on another side of town. All the firefighting force was shifted to control this new threat. As this fire was coming under control, a gale sprang up, whipping the first fire back into the attack, setting fire to the planing mill, from which sparks spread across town. The flames ripped into Cochrane on a hurricane of wind, blowing the panicked citizens to their knees as they fought their way to the lake in the growing gloom of smoke. Abruptly, a heavy deluge of rain came down, in time to save half the town.

With express train speed, a second branch of the extensive fire front engulfed the town of Matheson by mid-afternoon, felling people as they fled along the roads towards the river. The town was gone in about 15 minutes. Areas of cleared land were piled with the dried-out pine slash from the trees that had been removed. The dried coniferous branches burned explosively, like gasoline, and on a large scale, created a blowtorch that seared everything in its path in an instant. It snuffed out the lives of the running or hiding citizens of the settlements along the Temiskaming and Northern Ontario Railway, the principal communication between towns. The oxygenless, superheated gases of combustion simply rolled over them and pushed into their refuges, a condition very similar to the wartime firestorm of Hamburg.

All the larger places were on rail lines, and some escaped by train from these fires. Some would not go, with the strange reluctance to accept the clear warnings of recurring danger, and they died. Train crews managed to collect many people and pull out, often just ahead of the flames, reaching safety in Swastika and Englehart with some of the cars on fire.

All these frontier towns were wood and often congested. Cobalt, a mining boomtown of 3,000 at the time, fitted this description, was rebuilt the same way, and eventually suffered two more major conflagrations, the latest in 1977. The 1916 fire was named after Matheson. It had burned through parts of 49 townships, the worst in Canadian history. Somewhere in the vicinity of four to five hundred lives were lost.

By 1920 the use of fire watchtowers was coming in, together with the rudiments of organized reaction in firefighting operations by the provinces. The use of aeroplanes for aerial surveillance and transport of fire crews and their supplies was being considered. Towns were now more diligent about clearing the bush around their perimeters.

Late in September 1922 it was dry in Northern Ontario, but the fire wardens had packed it in for the season in the interest of the taxpayers. Nevertheless, the customary lumbering operations and farm clearing were still going on around the Temiskaming country. With these went the burning of slash piles. At the end of the month, rising winds whipped a number of these into the surrounding forest, and they grew together in a broad-fronted fire that picked up speed and, with ideal conditions of weather and fuel, bred its own winds of firestorm strength. Under the conditions we call a firestorm, what little or great natural wind there is feeds and gives impetus to the rising thermal column of hot gases from the fire. The column is bent forward at an angle by the natural wind, carrying smoke, heat and flying brands up and out over the leading edge of the actual fire front for miles. This happens in intense, large-scale fires both in forests and in cities. In the latter, there is highly flammable building construction, as in Saint John in 1877, or a large number of simultaneous fires being set, as in wartime Hamburg and Tokyo.

In the North in those days of little fire control, the presence of heavy smoke in the air was largely ignored by people in the towns. On October 1, 1922, the growing fire had been burning for days and progressed through several townships, covering hundreds of square miles. The smoke was blowing in the direction of large towns like Haileybury and Cobalt without causing any alarm.

It was not until the smoke overhead became very heavy and the glare could be seen at night on October 3, that there was any concern. Still, little was done. Feeding on slash and rotted windfalls, the fire built up into a racing firestorm that bore down on the hamlets along the railway northwest of Haileybury. Charleton, Englehart, Earlton and other communities were burned to the ground. The citizens of New Liskeard fled into Lake Temiskaming or tried to wet down their roofs with lake water. Luckily, the fire juggernaut missed its target, cutting through only part of the town. Haileybury was not so lucky.

At 3 p.m. on the 4th, with the fire within sight of Haileybury and the roar of its furious advance quite audible, the fire department went out to the edge of town, apparently with a steam fire engine, and prepared a stand, falling ash and increasing heat building around them. It seems they had faith in the cleared perimeter and anticipated a mild exposure problem.

Ruins mark the swath of destruction caused by the great Haileybury fire that came storming out of the surrounding forest. – *PAC-C44129*

Suddenly, the unexpected wall of flame burst through the screen of heavy smoke, consuming everything in its path without hesitation, racing toward the thin line of firemen. Knowing they were powerless, the fire crews fled back into town. The mill whistles blew and the citizens raced for the safety of the lake. The livelier ones made it. Others were incinerated in the streets or their homes. A big, prosperous town of 10,000 souls was wiped off the map in a couple of hours. Hotels, mills, cathedral, hospital, banks, schools, stores, railway terminals, all had disappeared.

A warning was carried by those fleeing early in cars to nearby Cobalt. Here again, there was little or no time, and the people fled into the lake. North Cobalt was flattened in minutes, a thousand residents racing towards Cobalt. The miners of the bigger town seem to have been made of stern stuff. With the technology they knew, they went to work with dynamite, picks and shovels to create firebreaks. Hoselines were also laid out. Blackened and burned, they succeeded in deflecting the edge of the big fire that, fortunately, just brushed the town. The huge Haileybury fire then turned back upon itself and burned out, having destroyed 2,000 square miles of forest and numerous communities and mine sites. As usual in such events, the life loss was not known, but was probably well over 100. Property loss was set at $8,000,000.

The Haileybury catastrophe finally moved the people of Ontario, who moved the politicians. Relief trains were rushed to the North and most provincial governments began to work to improve forest fire control. The aeroplane was becoming more effective and was brought into the picture for patrols, mapping fires, carrying firefighting crews and, eventually, waterbombing. Government firefighting organizations were established and specialized pumps, hose and tools developed. A light backpack pump was produced from an outboard motor.

Another fast-moving forest fire occurred in the Fort Frances area in Dance township in September and October 1938. It was driven by high winds, but there were few communities in its path. By this time, there was organized forest firefighting, utilizing aircraft, not that the forces available would have been strong enough to affect this fire in its course. After World War II, severe fires, mainly in unsettled areas, occurred in 1947 and 1948. Experience was now being accumulated to fully develop waterbombing, the smoke jumper firefighter, infrared mapping of fires, and other techniques to make the forest towns less vulnerable.

Institutional Tragedies in Quebec

A number of heavy life losses by fire occurred in the Province of Quebec in the first quarter of the century, mainly in religious institutions. Many of these structures had been built in the predominantly Roman Catholic province over a long period when construction practices showed little consideration of fire protection; building codes were weak and enforcement inadequate. They were, sadly, so often cases of humanitarian works that boomeranged because of ignorance in fundamentals of fire safety. It started with the Longue Pointe Asylum, a huge mental institution on Montreal Island, which burned for two days in 1890, with a loss of over 100 lives. The conditions of wide-open stairs and corridors, combustible construction and flammable finishes, together with ignorance of the potential for disaster would become a familiar pattern.

The Grey Nunnery on Dorchester Street in Montreal was a fine example of the huge institution, built over many years to serve a labour of dedication to people in need, but in total innocence of the need for greater attention to their safety from fire. On February 14, 1918, the five-storey building of brick and stone, a block long, with a combustible wooden interior, was occupied in part by 200 convalescent soldiers. In the whole building there were about 1,000 people: nursing sisters, nuns, aged and crippled people, and on the fifth floor of the west wing, was a creche for homeless infants.

Fire broke out in the early evening on the top floor near the central tower and spread

A makeshift fire escape can be seen added on the side of the huge Grey Nuns' institution in Montreal, but it does not reach the burned out attic where 53 infants died in the tragic fire of February, 1918.

rapidly through the combustible open interior. When the first fire companies arrived, the ambulatory soldiers were already at work, handing the tiny children down the fire escapes at great risk to themselves, the fire burning over their heads. Supplementary exterior fire escapes had recently been added, but not from the fifth floor. This whole top floor was soon ablaze. The firemen also dashed in to seize up infants from their cribs, some only a few days old. Only on direct orders from the fire chief in charge would the soldiers stop their work, some being seriously injured by falling debris. The fifth floor under the pitched roof was gutted before control was gained. Everyone escaped except for 53 helpless infants, their bodies found in the debris, victims of the merciless flames and gases.

A heavy life loss by fire in the Hospice St. Charles in the City of Quebec on December 14, 1927, was a similar institutional disaster. A large stone building of four and a half storeys, with combustible interior and open stairways, allowed fire to spread rapidly up from the basement to trap 37 people, most of them little girls between 5 and 9 years of age. At the time of the fire there were 371 children sleeping in the various dormitories and about 20 sisters in charge.

The hospice was an orphanage and boarding school. It had three wide stairways and two outside iron fire escapes. At 10:30 p.m., when all were in bed, the crackling of burning wood roused a nun, who immediately gave the alarm. The sisters carried out a disciplined and orderly exit of the children according to plan, but it was too late and many had to be taken down ladders or jump from windows into life nets held by firemen. The first alarm was recorded from a street box by the Quebec Fire Department at 10:48 p.m., sounded by a passerby.

Second and general alarms were quickly ordered by the deputy chief, the firefighters having difficulty with snowbanks and a high iron fence around the grounds. Most of the lives lost were by asphyxiation, and there was evidence that some children would not go through the smoke, but returned to their beds. A coroner's jury recommended that automatic sprinklers be installed in all institutions of this type, but it fell on deaf ears.

Another, almost nineteenth century brick-joisted institution, built in 1901, was the two- and four-storey Sacred Heart College at St. Hyacinthe. The residential school had the typical lack of interior cutoffs and was fitted with wholly unsatisfactory external fire escapes. Fire protection consisted of manual equipment, fire alarm, standpipe hose, and hand extinguishers.

In this case, fire followed an apparent coal gas explosion in a furnace room at about 2 a.m. on January 18, 1938. The temperature was -8 degrees Fahrenheit outside, and the five furnaces were being pushed. The college was occupied by 96 boys of 11 to 18 years, sleeping in dormitories on the third and fourth floors, while 26 priests and five employees slept on the ground floor. The interior of the structure was very quickly enveloped in flames.

Awakened by the explosion, the occupants were confronted by smoke and flames coming out of the hot-air heating ducts. Fire escapes and stairs were quickly blocked by flames. Some made their way out a chute-type fire escape, others jumped into snowbanks below their windows. Priests directed 43 younger boys to the roof of the two-storey section, where they were too frightened to jump to the snow below; and some remained on the roof until it collapsed soon after.

The inadequate town fire department was summoned by a passerby at 2:03 a.m., but could do little except raise a few ladders. Bolted doors could not be entered from the outside. A coroner's jury recommended putting dormitories on lower floors in such institutions, but the practice of putting incapacitated people and sleeping quarters on upper floors continued, not only in Quebec. The Dominion Fire Commissioner reported that in the period 1922 to 1938, 342 fatalities occurred in the eight largest of 4,200 school fires in Canada, and 90 percent of the life loss in institutions resulted from flammable structural conditions.

A dozen years later, in 1951, a similar institution, the Hospice St. Cunegonde in Montreal, burned in midday, with the loss of 35 lives, mostly aged people resident in wards on the fifth floor. Once again there was wringing of hands and promises of strict fire protection regulations on the part of the authorities, but we so quickly forget.

Not always have these institutional fires resulted in disaster, in spite of their vulnerability. In Baie St. Paul, on May 11, 1923, at a 2 a.m. fire in the St. Anne Hospital for the Insane, 250 male patients marched down the stairs in an orderly fashion without a casualty, while the building suffered a heavy loss. In Montreal, in 1925, 60 pupils in the Deaf and Dumb Institute evacuated their building with calm discipline as the roof blazed. Similarly, in Ottawa a year later, 53 young children filed quickly out of the blazing Protestant Orphans' Home. Some 2,000 mental patients were evacuated from the St. Michel l'Archange institution in Giffard, Quebec, in zero weather, in February 1939. The Quebec Fire Department fought the fire with 100 men, although it was outside of the city. It was a great problem to find alternate accommodation after this half-million-dollar fire.

These disasters in institutions, like others even up to the present, have been the result of both indifference and ignorance. Structural protection measures against rapid fire spread have long been known, and the availability of the automatic sprinkler system could have virtually guaranteed life safety for the occupants, even in the poorly-built structures.

Criminal Negligence at the Laurier Palace

The men of Montreal's No. 13 Fire Station, on Notre Dame Street East, had their quiet Sunday afternoon of January 9, 1927, interrupted by a "still alarm", that is, someone came into the station to report a fire in person. The fire was right across the street. It was in the Laurier Palace Theatre, and the Sunday matinee was on.

Going into action with alacrity, the firefighters found the patrons pouring out of the ground-floor entrance and some standing on the marquee at the balcony level of the 800- to 900-seat motion picture house. They commenced fighting the fire that was spreading from a concealed space under the balcony.

Then they made a grisly discovery. To their horror, the firefighters found one of the two narrow stairways coming down from the balcony was blocked solidly full with human bodies. They were all children and were wedged so tightly that the firefighters could not pull any of them out.

When smoke had first begun to rise in the crowded auditorium, there must have been a rush for the exits. The two balcony stairs were steep, enclosed by walls, had a right-angle turn part way down and narrowed toward the bottom. At the ground floor was a doorway only 37½ inches wide. Many of the children, who made up most of the audience, escaped from one stair without difficulty, but on the other side someone stumbled. There were no hand railings, and the downward rush resulted in a deepening pile of young bodies on the small stairway landing until they were eight deep.

This all happened within two minutes of the appearance of the fire. Then there was a smoke explosion that drove smoke into the stairway. There were 78 deaths in that jam, 52 by smoke asphyxiation, 25 crushed to death and one burned. Only one of the victims was over 16 years of age.

The firefighters quickly extinguished the fire and took up the painful task of cutting away the stairway from underneath to extract the bodies. A few at the top were brought out alive. Needless to say, the design of the movie theatre violated fundamental rules of exit planning, to say nothing of the ordinances of the city. The disaster could have been worse. The Laurier Palace was of very flammable construction, having a wood interior with wood sheathing and plaster on wood lath. The theatre was full, and there were standees at the rear and in the aisles. The probable fire cause was concluded to have been careless smoking.

It is ironic that this deathtrap existed right across the street from a fire station, constantly manned by a force of men paid for and dedicated to saving lives from fire. However, the preoccupation of the day in most fire departments was on fighting losses by fires after they had broken out and, to a lesser extent it still is.

The principal result of the Laurier Palace disaster was the absolute prohibition of anyone under the age of 16 from attending any motion picture theatre in the Province of Quebec. Until television, Quebec children simply did not see the movies.

An Old Hotel

Halifax, which was soon to become Canada's overcrowded "front line" city in World War II, lost one of its big, older hotels during 1939 in one of the country's worst hotel life-loss fires. The Queen Hotel was an accumulation of sections built between 1849 and 1908. The latest was reinforced concrete, the others brick with wooden interiors, and unprotected doorways between them made them all one fire area. Elevator and stair shafts were not cut off from smoke or fire spread by walls and doors. There were no sprinklers, alarm systems or watchman service, and the outside fire escapes were unsuitable. The main parts of the hotel were five storeys.

The conditions were right for disaster when, on March 2, 1939, a little after 6 a.m., the night clerk discovered that fire had broken out in the oldest section. With no alarm system, he had to start calling each guest room individually on the old-style switchboard. He had time to make two phone calls when smoke made it necessary to abandon his desk. There were 78 registered guests and 30 employees in the building. It is almost surprising, in the circumstances, that many escaped. At least 28 people lost their lives and 19 were seriously injured.

When the fire department arrived from its Bedford Row headquarters, only a block away, there were people jumping from the windows, inside escape having quickly been cut off. Many were taken down ladders or jumped into a fire department life net as limited manpower struggled valiantly to cope with the urgent situation. A maze of overhead wires hampered work with the one available aerial ladder. Firefighting could not commence until rescue work was finished, and the hotel complex was destroyed, along with two small adjoining buildings in the crowded downtown area.

The Queen Hotel fire was a turning point for fire protection in Halifax. The shocking revelation of conditions in the hotel and the effect of a decade of depression on the fire department were taken to heart. The fire department was reequipped and a high rate of sprinkler protection in buildings is now evident in this city.

There have been many smaller life losses in the older commercial hotels across the country, built with serious fire safety flaws in an age of weak codes and enforcement. The Barry Hotel in Saskatoon at least had concrete floors, but an otherwise combustible interior and open stairways. The alarm system operated on batteries, but these were dead at the time. The Saskatoon Fire Department rescued many of the 200 people in the hotel with ladders, as flames blocked the main stairway, but ten men and a woman were asphyxiated and 17 were injured.

(XX) Between the Wars

The Commission on Conservation

In 1909 the Canadian government passed a Conservation Act establishing a Commission on Conservation made up of a number of members representing prominent business, professional and university officials, as well as provincial cabinet ministers. Sir Clifford Sifton, K.C.M.G., was chairman. The commission was asked by the Canadian Manufacturers' Association, the Board of Trade and others to investigate the problem of needless fire waste. The result was a detailed study by J. Grove Smith in 1918, *Fire Waste in Canada*, covering the fields of building construction, fire prevention laws and education, and firefighting.

Not a great deal seems to have come out of the conclusions of the study, other than that it may have influenced the provinces that did not yet have a fire marshal or fire commissioner to appoint these officials. Mr. Smith, an engineer, became the first Dominion Fire Commissioner, charged with collecting fire statistics, promoting fire prevention generally, and seeing to fire safety matters within federal jurisdictions.

The conclusions of Smith's report covered the need for federal and provincial governmental action on fire safety education, fire research, enactment and enforcement of adequate building and fire prevention codes, reduction of arson and the cooperation of all elements in the struggle against fire waste. All of these were closely parallel to the findings of a second national study published 62 years later, in 1980, by the National Research Council of Canada, suggesting that there had not been very much fundamental change in the attitude of Canadians or their governments toward fire loss in the interim.

First Provincial Fire Commissioners and Marshals

At the time of the Grove Smith Report, three provinces had been engaged in the activity of fire prevention promotion and control through appointment of officials under Fire Commissioners' or Fire Marshals' acts, but there was no large-scale activity. Manitoba had been first, establishing a Fire Commissioner's Office in 1876, with an act empowering him to carry out fire safety enforcement, but little is known of any activity until 1917, when the job was held by the Superintendent of Insurance. Saskatchewan adopted a fire prevention act in 1912, the Superintendent of Insurance holding the position from 1917 until 1948. In 1916 a revised act made municipal fire chiefs local assistants to the commissioner, the practice almost universal today.

Ontario appointed its first fire marshal in 1916. The first to hold the office was E.P. Heaton, who, by 1918, had a budget of $50,000, indicating a fair-sized programme. While Alberta had passed an act in 1919, no official had been appointed. British Columbia had no act until 1921, though there were some legislative provisions for fire prevention inspection in existence in other laws.

Quebec had long had an act to set up regional or city fire commissioners with powers to hold enquiries, but only in the cities of Montreal and Quebec were there any commissioners operating. In the Maritimes nothing was being done in fire prevention on a provincial basis until 1920 and 1921 when Nova Scotia and New Brunswick appointed fire prevention boards, which in turn named the fire marshals up until 1940 and 1944, respectively. P.E.I. set up a fire marshal's office in 1919 in collaboration with the local board of insurance underwriters. The Association of Canadian Fire Marshals and Fire Commissioners was established as a forum for these officials in 1921, meeting jointly with the Dominion Fire Prevention Association until 1950.

This new involvement of provincial governments stressed fire prevention education, inspections and investigations in noteworthy cases, and the beginning of record-keeping on fire losses in these early days, but real help or guidance to the municipalities in the fire control job was a long way off.

The legacy of combustibility in Canadian cities and towns was of great concern between the wars. Here is a smoky one in Charlottetown. The horse-drawn reel has just laid an additional line of hose.

Federal Involvement

An attempt to organize fire prevention efforts in Canada on a national level began in 1912. E.P. Heaton, who was then insurance manager for the Canadian Manufacturers'' Association, addressed the annual meeting of the National Fire Protection Association (N.F.P.A.) in the United States on fire waste in Canada and appealed for international cooperation. The outcome was the setting up of a Canadian Committee of the N.F.P.A. The following year, an Ontario Fire Prevention League (O.F.P.L.) was formed and the bill for the appointment of an Ontario Fire Marshal introduced in the Ontario Legislature. Mr. Heaton received this appointment, and the O.F.P.L. was designated to help him with his programme.

With a war on and losses of critical goods occurring, attempts to step up fire prevention activity in the country were promoted by such organizations as the Canadian Manufacturers' Association, Canadian Fire Underwriters Association, the Trades and Labour Congress, boards of trade, chambers of commerce and municipal councils. Seeking Dominion government action, they ran up against the constitutional provision that fire protection for the public was a provincial jurisdiction. Although there was a suggestion made that the government should take measures through the War Measures Act, then in effect, a request was put forward to the Commission on Conservation, asking it to "stimulate and coordinate" the provinces in fire prevention activity.

The commission proposed that a competent engineer from its Town Planning Branch be appointed to study the question. J. Grove Smith was named to do this, and a Fire Prevention Branch set up in 1916. Smith produced his 1918 report and the next year a Dominion Fire Prevention Committee was set up to advise the government, in particular, the Fire Prevention Branch. This committee became the Dominion Fire Prevention Association (D.F.P.A.) with representatives from all areas of business and industry. Grove Smith was appointed first Dominion Fire Commissioner, to carry out educational and advisory work in fire prevention and to act as executive officer of the association. The D.F.P.A. continued in existence, apparently with slowly declining influence upon the government, until it expired in the 1950s.

Grove Smith died in 1939 and was succeeded by another engineer, W.L. Clairmont. After the appointment of C.A. Thomson to the position in 1950, renewed interest in the protection of growing federal property against fire resulted in expansion of the Dominion Fire Commissioner's Office. It became a fire protection engineering organization that developed and enforced standards in this field, substantially upgrading federal buildings. Excluded from this programme were some crown corporations and the Department of National Defense, the latter having its own fire marshal, successor to the three wartime fire marshals of the army, navy and air force.

The Fire Chiefs' Associations

The National Association of Fire Engineers, founded in 1873, was the first major association of municipal fire department heads in the United States. Chief Benoit of Montreal and a couple of other Canadian chiefs attended their meetings, too. Benoit, in fact, rose to be elected president of the association and brought it to Montreal for its annual conference in 1894. It thereupon became the International Association of Fire Engineers, (later Fire Chiefs as the term engineer fell into disuse) and, finally, set up regional divisions, with a Canadian division.

The Canadian division was in the form of the Dominion Association of Fire Chiefs (D.A.F.C.), an association founded in 1908. The D.A.F.C. became the Canadian Association of Fire Chiefs and broke from the I.A.F.C. some time later after World War II.

The Depression

With the Depression many things ground to a halt, but with firefighting an essential service, it appears the thirties brought less in the way of reductions in protection than in postponement of improvements and expansion. Sales of fire apparatus slowed, and the manufacturers featured economy models. Capital projects for water supply and fire stations were likely less urgent as building generally slowed down. On the other hand, the City of Montreal, under its colourful and perennial mayor, Camillien Houde, launched make-work projects by installing municipal services in the open fields on the outskirts. Streets, water mains, hydrants and even fire stations were put in where no buildings would appear to fill in the new city blocks for some time to come. Fire stations 47 and 48 thus remained unused for some years.

There were delays in improvement of working conditions, though hours were reduced in Calgary. This came, along with a pay cut, by agreement, in order to eliminate layoffs. Each man took a half day off per week. Vancouver closed two stations completely. Postponement of reequipping and expansion generally meant delay on through to the end of World War II, as material shortages quickly succeeded the cash shortage.

On the Waterfront

The greatest waterfront fire to occur in Canada started on the morning of June 22, 1931, in pier transit shed No. 5, on the West Side Docks at Saint John, N.B. The fire raced rapidly through the large unbroken areas of the 12 wooden pier sheds, along the grain conveyor galleries between the sheds and the elevators to other piers, and across the pier decks, all of it wooden construction. The transit freight sheds ranged from 23,000 to 50,000 square feet in area.

The conflagration destroyed everything southward along the port's principal docks for ocean shipping to the last pier and shed, No. 17, brand new and not quite finished, at the outer end of the harbour. The docks were the property of the Saint John Harbour Commission. Grain conveyors, considerable rolling stock of the C.P.R., and a 600,000-bushel wooden elevator were lost, as were several dwellings and the government immigration shed. This last facility had been equipped with automatic sprinklers, but was swallowed up in the maelstrom without being noticed.

The other buildings had only some hand extinguishers, standpipe hydrants and hosereels. The fire was discovered at the outer end of No. 5 Shed at about 11 a.m. The steamer *Gunderson* was unloading bananas and there was considerable hay used in their packing, now lying about in the shed. After attempts to extinguish the flames, probably started by a carelessly discarded cigarette, the fire department was called at 11:15 a.m. The call was answered by No. 6 Pumper, the only fire company in the city located on the west side of the harbour. No. 6 Fire Station had only just recently been motorized, so the big LaFrance pumper was very new. The crew drove down the pier and through the long shed, thinking there was a very small fire to handle. When they were far into the shed, smoke and flames came bursting through the shed floor, having spread below the deck. The new pumper was enveloped, and the crew barely escaped, leaving their pride and joy behind.

By 11:30 the entire shed was a mass of flames, and the fire was spreading to the neighbouring sheds and racing along the dust-laden grain galleries high in the air above. The *Gunderson* was quickly moved clear, but the old laid-up bay ferry *Empress* was burned. Fire Chief L.W. Vaughan sounded a rare general alarm at 11:40, bringing five more pumper and hose companies and a ladder company, all he had except one, around the Reversing Falls Bridge. The volunteer pumper company from adjoining Beaconsfield also responded.

By this time the conflagration was well out of the control of such a force. There was no fireboat. The three major pier sheds — 5, 6 and 7 — were a raging mass of flames, fanned by a brisk northwest wind. About noon, fire crossed the slip to sheds 2, 3 and 4. In another hour the big wooden grain elevator, sheds 14, 15 and 16, seven houses on Protection Street, and some 125 boxcars were involved.

Flames travelled through the wooden grain conveyor galleries, but the fire was kept out of the new million-bushel concrete and steel elevator. Efforts to dynamite the galleries' supports were only partially successful. The arrival of the east-side fire companies, though prompt, was not in time to accomplish anything, except to take up a perimeter position along Union Street at the edge of the waterfront.

In a mere three hours, 11 ocean shipping berths and their shore facilities were consumed by this fire, a straight-line distance of about one mile. It was a memorable experience for a young visiting volunteer fireman from North Sydney, Ed Willis, who pitched in with the locals, gaining some of the experience that would serve him as chief of his own department many years later.

The loss in this devastating fire was set at about $5,000,000, only a fraction covered by insurance. No new lessons were learned. It had been predicted — which took little expertise — by a fire protection engineer for the New Brunswick Board of Fire Underwriters, and a proposal for sprinkler protection, as well as a fireboat, was made. No action was taken.

A mile of wooden dock facilities goes up in smoke in an afternoon at West St. John, N.B., in June, 1931, in an easily predicted calamity that crippled this port.

Long wooden piers and undivided, unprotected sheds on them very rapidly become a mass of seething flames once fire gets a start. The strongest fire departments are helpless without built-in protective measures.

The Cymbeline

The oil tanker *Cymbeline* was high and dry in the floating drydock at Vickers shipbuilding yard in Montreal harbour on June 13, 1932, undergoing repairs as the result of going aground. The oil from her damaged tanks had been transferred into ballast tanks in the steel walls and hull of the floating dock, then the tanker was steamed clean and tested free of gas in accordance with safe practice.

A night crew was rushing the repairs when, at about 3:30 a.m., two of the bottom tanks of the drydock blew up, showering blazing oil over the drydock and tanker where the men were working. The men on the deck and inside the tanker struggled to escape, their clothing oil-soaked and in flames. Some succeeded by jumping into the harbour, others burned to death where they stood. Members of the crew of the tanker were thrown from their bunks in the forecastle by the explosion, but managed to escape. The explosion resulted in the deaths of 26 men.

An hour later Montreal Fire Chief Raoul Gauthier and some of his men were standing on the drydock wall directing streams of water on the now burning superstructure of the ship. A second severe explosion then occurred, this one in one of the sidewall tanks of the dock, directly under the firefighters. It killed Chief Gauthier and three firefighters. It was surmised that vapour from the tanks used for temporary storage had been created and then ignited by heat from the rivetting forges or acetylene torches being operated on the floor of the dock. Gauthier had been a dashing heroic figure as a fire chief, and his loss was felt by the citizens.

Vancouver's Railway Pier Spectaculars

The port of Vancouver suffered two very similar, spectacular pier fires in the thirties. The first, destroying the more than 1,800-feet-long Canadian National Steamships ocean and coastal terminal on August 10, 1930, should have given a hint to the Saint John Harbour Commissioners. The C.N. pier was of creosoted wood piling, with two one-storey freight transit sheds of 790-foot length, and at the head, a two-storey office and passenger building. The structures were of wood with some unprotected steel trusses in the roofs.

The facility had just been reconstructed and was in use. While there was an automatic sprinkler system, due to some poor scheduling it had not been connected to the water mains yet. There were none of the modern forms of structural fire cutoffs, the sheds being wide open from end to end, as was the pier substructure. As in the Saint John fire, a small blaze was discovered at the extreme outer end of the pier at 3:40 p.m. Two watchmen tried to fight the fire with the inappropriate carbon tetrachloride extinguishers provided.

The Vancouver Fire Department received the alarm at 3:52 p.m. by telephone. On the arrival of District Chief Eaton and the first alarm companies, about 250 feet of the outer end of the west side of the pier was heavily involved in smoke and flames. Eaton ordered two lines of hose laid from hydrants at the head of the pier. These were 1,250 and 2,000 feet long, an excessive length. The fire advanced so fast that the firefighters were forced to retire before the water arrived, abandoning their hoselines and only just saving their trucks.

A number of the freight loading doors were open and the draft aided in the spread of the flames. There was a division wall between the two long, narrow sheds, where they were joined at the outer end. This apparently failed and both sheds became involved. Deputy Chief Loftus arrived and took command, putting in a third alarm at 4:01. He abandoned the long pier sheds to the roaring creosote-fed inferno and concentrated his forces on a successful defense line where a bulk oil depot was closely exposed at the shore end.

The loss of a large and important dock facility, at a cost of $1,050,000, had become an accomplished fact in about a quarter of an hour, during which fire had travelled a thousand feet. Again as in Saint John, there was no fireboat service available. Such a vessel would have been able to bring to bear the massive power of its streams to match the large-area fire and to hit the fire under the pier deck. There was a fireboat on Vancouver's second harbour, False Creek, but it could not have rounded Stanley Park in time to help.

Two fireboats had been recommended for Burrard Inlet, the main harbour, by the B.C. Underwriters Association, along with more water mains, hydrants, alarm boxes and structural protection. Action had not been forthcoming due to what would be a long-lasting difference of opinion on responsibility for harbour protection between the various port cities and the federal government. Burrard Inlet would not have a real fireboat until 1951, when the city took on the responsibility. Except for Toronto and, briefly, Quebec, other major Canadian ports would never have other than makeshift or naval fireboats available except in wartime.

The second harbour pier fire in Vancouver was a virtual replay of the C.N. fire, but further west in the rival C.P.R.'s Pier D. On July 27, 1938, about 1:30 p.m., a fire started at the outer end of the wooden open pier, racing with astonishing speed up to the head of the pier, wiping out the 900-foot-long freight and passenger service superstructure.

The fire took about an hour to reach the head of the pier, travelling above and below the deck of heavy timber. The transverse bulkheads served as only a momentary check, the flames sweeping around the edge of the pier aprons on both sides to envelope the pier shed from the outside.

The Paper Mill Blockpiles

One special type of fire had the odd feature of going on so long and so slowly that, once underway, quite different and unhurried methods were often used. This was the paper mill blockpile fire. Coal mines, peat bogs and town dumps are about the only comparable long-term fire situations.

Before pulp and paper mills generally converted to large storage of raw wood material in the form of chips, the standard method was to accumulate mountainous stockpiles of softwood, mainly spruce, peeled and in four-foot lengths. This was the stock that fed the pulp- and paper-making process, and amounted to as much as a hundred-thousand cords in a pile. Until some very costly fires occurred in them in the late 1920s and early 1930s, fire protection was fairly casual.

The greatest blockpile fire took place at a paper mill at Port Alfred, Quebec, in a pile containing over 200,000 cords, an abnormally large one, about three city blocks long. The fire started about 3 p.m. on April 29, 1932, and was considered out about 6 a.m. on May 11. The fire ate its way through most of the 1,700-foot by 700-foot pile that was as high as a ten-storey building. It burrowed deep into the interior of the wood, where hosestreams could not reach it.

These conditions resulted in a firefighting operation that required marshalling of large pumps from far and wide, setting up temporary electrical substations to power them, bricking up windows in the paper mill to protect it, and laying temporary water mains. Ultimately, a tug was called from far off Quebec City to tow the huge suction-type dredge *Montcalm* to the scene, arriving on May 4.

With the mill's own powerful pumps, numerous temporary stationary pumps and piping, the huge capacity of the suction dredge was added. It fed water from its position out in the Saguenay River through 2,000 feet of 24-inch-diameter pipeline laid up the side of the pile. This delivered ten to 12 thousand gallons of water per minute on the pile, possibly the largest single firefighting stream ever pumped on a fire. The total rate of application on the fire peaked on May 9, when just under 40 million gallons were delivered on the pile.

Other pumping records were established on a smaller scale. A LaFrance fire department pumper from Chicoutimi arrived on the scene early and pumped continuously for 96 hours before being stopped to replace a blown manifold gasket. A Bickle pumper from Quebec

Fire Department also operated 96 hours before developing a hot bearing. Quebec sent along a 62-year-old steam fire engine, as well. This machine pumped 76 hours and was withdrawn due to burned-out grates and boiler tubes.

The great fire was stopped and some 35,000 cords (out of 210,000) were saved, along with the paper mill toward which the fire was heading. Over 1,000 workmen were engaged in fighting the fire. After this, the insurance interests saw to it that blockpiles were ringed with large monitor nozzles fed by high-capacity pumps.

(XXI) Fire Service in World War II

The War Effort

When Canada went to war again in September 1939, there was another exodus of ablebodied men from fire departments, career and volunteer, to the armed forces. Some of these soon turned up in the military fire departments set up to protect the many new bases being thrown up across the country. More than one new naval rating expecting to go to sea found himself collared at the Halifax dockyard and told he would be a firefighter; his previous experience had come to light and there was a pressing need for it. One of these was William Carson, now destined to win his medal fighting fires, not U-boats, as will be seen.

Most military camps of any size had a pumper and crew, supplemented by fire *picquets* from the military units stationed there. Larger establishments had several pieces of major apparatus, and the smallest had trailer pumps. Mutual assistance was arranged with nearby cities and towns.

Supplies of apparatus and equipment for municipal departments held up fairly well for a time, American production of components being available until their intensive mobilization after Pearl Harbor in December 1941. Following the example set in the British Auxiliary Fire Services, voluntary auxiliary fire companies were set up under the Civilian Protection Corps organization in key cities, notably the coastal ports. These companies drilled under the municipal departments, turning out on the periodic blackout drills. They manned trailer pumps of several sizes and proved useful when large fires taxed regular fire departments.

Industry wound up to an all-out multiple-shift production effort. Strained and makeshift facilities, inexperienced help and heavy concentrations of goods made for more and bigger fires. The load of handling them fell on weakened municipal fire departments with aging equipment. Thoughts of possible enemy sabotage often arose when a major fire hindered production or consumed essential materials, but none seems to have been substantiated.

One of the largest fires was the loss of an 1,800-foot-long pier shed on the Lachine Canal in Montreal, loaded with scarce and rationed commodities, on May 6, 1944. Drums of alcohol fuelled the flames that moved through the undivided structure so fast that some of the employees barely escaped. The Montreal Fire Department used 23 powerful streams of water on it, but had poor access since it was on a long pier and there was no fireboat. During the war, the Montreal department also handled a fire involving a group of tenements into which a heavy bomber crashed shortly after takeoff from Dorval Airport. Eight lives were lost in addition to the aircrew.

Data on fires in military establishments was not freely available in wartime for security reasons. This secrecy quite possibly was advantageous when there were goofs to be covered up, as well. When the National Fire Protection Association asked for details from Ottawa about a $400,000 hangar fire on one of the many Prairie air bases, it was advised that the fire was of a type that could only occur on an air base, that there was nothing that could have been done to minimize the loss and, therefore, no useful lessons could be learned of benefit to N.F.P.A. members. The unofficial story that leaked out was that a plane crashed

A new bomber heading out of Montreal for Britain during World War II crashed into this tenement district. A piece of the wing gives the evidence as Montreal firefighters search for victims.

The old Toronto harbour tug Rouille became the mainstay of marine firefighting operations in crowded Halifax Harbour during the war. Here she is subduing the fire that destroyed the Dartmouth ferry Governor Cornwallis, December 22, 1944.

into a combustible hangar that had no sprinklers or other suitable protection. The water supply was only that available through a long, dead-end six-inch main, and a booster pump had been installed to improve the supply but never connected up. Wrenches did not fit the hydrants or hose couplings, and there were other deficiencies. There was apparently no structural fire truck on the base. No lessons, indeed!

In 1943, in a one-month period, there were three serious fires in the large permanent military base of Camp Borden in Ontario. Two were in air force hangars and the third was a large loss in the destruction of an ordnance workshop. The last was a combustible building lacking in adequate fire protection. The loss was estimated at between $2-5 million.

Early in the war, the military was hard pressed to find accommodation for rapidly expanding personnel, taking over all kinds of buildings until new camps and bases could be built. The local exhibition grounds and buildings, often on military land, were logical for this, as they lay vacant most of the year. In Saint John and Fredericton, the huge old wood and glass exhibition halls, remotely imitative of the Crystal Palace style, became drafty, uncomfortable accommodation for fresh recruits. Heads nodded knowingly when both these buildings burned down in short, spectacular blazes, and the soldiers had to be moved to other quarters. The standard army H-hut barracks, built of wood and covered in heavy tar paper, were reputed to be capable of burning flat in about 15 minutes, but this was never officially confirmed.

Halifax was the key naval base and a concentration point for the convoys of merchant ships crossing the Atlantic. Real war came closest with the activity of German U-boats on the Atlantic seaboard and in the Gulf of St. Lawrence. The port cities were dimmed by turning off neon signs and other unnecessary outside lighting.

Casualties from the sea war, half-sinking freighters with torpedo holes and damaged vessels leaking fuel and flammable cargoes, all came limping into ports like Halifax, Sydney and St. John's. The pilings under the old wooden docks of Halifax became coated with flammable oil deposits. With their crowded wooden buildings, they presented a worrying hazard, exposing the downtown core.

Halifax was badly overcrowded during the war, as the main base for supply of men and material to Britain. As a result, the water supply, among other services, was taxed and weakened. One major church fire frustrated the fire department due to insufficiency of water from the hydrants, as did the burning of a large lumberyard, despite the navy's help.

The Fireboat War at Halifax

For Haligonians the burden of overcrowding by an influx of strangers from across the country brought back the memory of the 1917 *Mont Blanc* holocaust. The fear of a repetition was not unreasonable as long as processions of ships filed past the downtown district, between the ocean and the big assembly anchorage for convoys in Bedford Basin. Many of these anonymous grey freighters and tankers contained cargoes of ammunition and gasoline.

Usually unknown to the citizens, a doughty collection of civilian and naval firefighters risked their lives in a number of touch-and-go close shaves, as the inevitable fires occurred aboard. Ships were being worked hard and many crew members were inexperienced.

When the war broke out, the Royal Canadian Navy had few ships and certainly no fireboats. The federal government had converted its only fireboat in the country to a pilot boat on the St. Lawrence River. With the outburst of activity by the navy and the setting up of the North Atlantic convoy system at Halifax, steps were taken to provide harbour fire protection and salvage. The navy acquired the standard two-wheel trailer pumps and could sling them aboard a harbour vessel or barge as needed.

The old 85-ton steam tug *Rouille*, of the Porter dredging fleet, was fitted out with three monitors, chemical foam and a centrifugal fire pump of 2,000 g.p.m. capacity. It was operated by the National Harbours Board and provided with a firefighting crew under Captain Donald Preston, by the Halifax Fire Department. *Rouille* went into service in July 1941 and served valiantly until August 1945, when it returned to Porter's.

A second fireboat was added in August 1943, when the old vessel *James Battle*, retired from the Detroit Fire Department in 1935, was refitted for service and manned by the N.H.B's. shore fire patrol as needed. It had six pumps and an output of 7,000 gallons per minute. Late in the war, one of three small 3,000 g.p.m. capacity fireboats, built on ville-class tug hulls, was commissioned in Halifax by the navy. The others went to St. John's and Esquimault. These boats continued to serve in these ports for many years after the war.

Alone on the job, *Rouille*'s first call was to a fire in the 20,000-ton Canadian Pacific liner *Duchess of Atholl* at Pier 2N, a four-hour, four-stream fire. She also fought the fire that destroyed the new Halifax-Dartmouth ferry *Governor Cornwallis*, and that aboard the American trooper *Wakefield* (the former liner *Manhattan*), which had come into port with a fire aboard. Captain Preston received a commendation from the U.S. naval observer for this crew's efforts.

The French freighter *Maurienne* suffered a serious fire at Pier 28 in February 1942, and *Rouille* fought this fire along with land companies. The Halifax Fire Department, under then Deputy Chief Fred MacGillivray, removed 400 depth charges from her cargo.

Halifax came near to a repetition of the *Mont Blanc* calamity twice in World War II. The first was the case of S.S. *Trongate*. This old 7,000-ton British freighter had been a problem, having been under repairs and lying in the harbour for some time. There had been a one-hour fire aboard on March 29, 1942. On April 9, she was at anchor with no steam in her boilers in the midst of a number of American naval vessels at the main harbour anchorage, right opposite the city, near George's Island.

Trongate was loaded with cargo, including 3,000 drums or 570 tons of toluol, 10,500 cases of 25-pounder cartridges, shells and small arms ammunition. This has been represented as equal to more explosive power than *Mont Blanc* carried, but reported figures on the amount vary in both cases.

Fire broke out aboard *Trongate* again at about 9:30 p.m. on the 9th, and the crew fought the fire in four and five holds for a time before calling for help. At 9:50, *Rouille*, with Captain Preston, and several tugs steamed out in response to frantic signalling by lamp. The captain of *Trongate* announced to all who could read code that he had TNT aboard, then that the fire was out of control and he was sending the crew ashore. Muffled explosions and belches of flame and smoke came from the stricken vessel as *Rouille*'s firefighters and a Foundation Company tug tried to fight the fire.

No progress was made over several hours, the firefighters withdrawing at 1:30 a.m. with the intention of scuttling the ship where she was. This proved unfeasible, and at 2:45 a.m. the nervous citizens of the two cities were awakened to the sound of gunfire. The Bangor-class minesweeper *Chedabucto* had loaded up with nonexplosive practice shells and was punching holes in the engineroom of *Trongate* along her waterline at point blank range, while burning pyrotechnics shot into the night sky from her victim. It took a long time, but was in time, as *Trongate* sank to the bottom of the harbour. Only her masts showed above water among the other ships for ferry riders to puzzle over next morning. The ill-fated ship stayed there until cut up in 1954, though her superstructure was removed as a navigational hazard.

The Volunteer Caper

The story of the handling of the fire that broke out in the American munitions carrier *Volunteer* has been a favourite and often told, but is required repeating here.

The elderly freighter of 10,000 tons was lying at anchor amid 80 ships in Bedford Basin, all awaiting departure of their convoy. The crew of *Volunteer* were required to stay aboard to safeguard their cargo. Unable to have shore leave, the bored captain and several officers were playing poker and evidently imbibing substantially during the night of November 2, 1943. At about 7:20 a.m. an inexperienced stoker set fire to the boiler room while firing up a boiler. With the officers incapacitated, there was confusion aboard as the fire spread, causing delay in calling for help. Firefighting by the crew was ineffectual, and the officers continued playing poker.

Eventually the captain sent out an SOS by wireless to the surprise of the Camperdown wireless station, as they identified it as coming from inland and a breach of wireless silence. The naval dockyard was notified and this put the problem in the lap of the King's harbour master. The *Volunteer*'s holds, to which the fire was now spreading, contained 1,800 tons of heavy howitzer shells, small arms ammunition, dynamite, more than 200 drums of magnesium, bagged phosphates and wheels of tobacco. At that, it was not unusual to understate the amounts of explosives aboard ships at that time.

Earlier, in March of that year, a new dockyard commander and King's harbour master had been assigned in Halifax. He was Captain Owen C.S. Robertson. At six feet seven inches, he was about the tallest man in the navy, an unorthodox operator in the navy scheme of things and a man of action in an emergency. The harbour master's responsibilities included the disposition and welfare of all vessels in the port. This included fire protection. According to Robertson, he worked out a good cooperative mutual aid deal with veteran Fire Chief Churchill of Halifax over a bottle of rum, and the two always worked well together in a variety of fire emergencies. A similar deal was made with Captain Cody of the Harbours Board.

Some of the naval firefighters of Halifax who handled potentially disastrous munitions fires during World War II. Included are the captain of the dockyard, (the tall one) later Commodore O.C.S. Robertson and the later fire commissioner of Saskatchewan, William Carson (second from right).

On this occasion, "Long Robbie" Robertson turned out with his fire chief, Lieutenant Spinney, Lieutenant Buxton, his Master Attendant; Stoker Petty Officer Carson and other navy firefighters, along with the fireboats *Rouille* and *James Battle* and their civy fire crews. The navy men boarded the burning *Volunteer* to find the crew had abandoned ship. Cargo was burning in No. 3 Hatch, the fire having communicated through from the boiler room. Robertson and his men went down into the hatch to reconnoitre and ventilate the fire, only to be blown right out again by an explosion. P.O. Bill Carson, a former civilian firefighter from Saskatchewan, and Stoker Shatford were injured and others gassed by the explosion. Robertson says he woke up with water being poured on him to revive him.

Carson carried on with the arduous operation, but Stoker George Shatford was hospitalized and later died from his injuries. Robertson and his men re-entered the holds to carry on the battle. A barrier of tobacco was built to fend off the small arms ammo that was flying about as they vented the fire around the magnesium drums. *Rouille*, with the Halifax

Fire Department crew, pumped foam into the boiler room. The other fireboat crew was about to cast off and retire as things heated up, but Robertson made them stay at gunpoint. Meanwhile, the officers of the *Volunteer* continued their card game in the saloon.

It soon became necessary to flood No. 3 Hold, as the fire intensified. Attempting to handle an acute emergency on another man's ship was a challenge. Altercations resulted while the ship's captain continued his card game and Robertson searched the chief officer's cabin to find the ship's cargo plan. Deciding *Volunteer* was a danger to the other vessels crowded in the anchorage, Captain Robertson called out a pilot and had the U.S. naval liaison officer come aboard. In the saloon of the burning ammunition ship, punctuated by small explosions, the captain was formally relieved of his command and *Volunteer* turned over to Robertson, a tall apparition coated in soot from his frantic work below.

Now he had dockyard men cut away the fouled anchor cable and, *Rouille* alongside pumping, harbour tugs towed the floating time bomb towards The Narrows and Halifax harbour under the direction of a cool pilot, civilian John Brackett. Down past the unsuspecting city they sailed, small arms ammunition popping off and the vessel listing drunkenly. Finally the expedition arrived at the scuttling ground at Mauger's Beach on McNab's Island. At 4 p.m. *Volunteer* was sunk to sit on the bottom at just the right depth to drown the fires, but with the deck high enough so she could be pumped out and refloated again.

The ship and most of the cargo were salvaged. Captain Robertson was awarded the George medal for stopping the otherwise inevitable disaster when the crew abandoned ship. British Empire medals were awarded to Carson and Brackett. The former rose to become dockyard fire chief and, back in civilian life, fire commissioner of Saskatchewan.

The Canadian Corps of Overseas Firefighters on review in Trafalgar Square, London. These career and volunteer firefighters from across the country arrived to help their British brothers after the main Blitz.

The Canadian Corps of Overseas Firefighters

Expecting the aerial bombing of their cities, with intent to burn them, the British government prepared a strong fire defense. Fragmented local fire brigades would not do, so all of the municipal fire services were combined into a single National Fire Service. To this service were added auxiliary firefighters, special apparatus, trailer pumps pulled by taxis, temporary water-main-laying units, and the logistics of mustering large forces over great distances. An organization of 100,000 people was put together. The Blitz resulted in a severe trial, many conflagrations occurring in London and other cities easily accessible to Germany's short-range bomber force.

With Britain suffering under this onslaught by fire, Prime Minister Mackenzie King of Canada suggested that a force of Canadian firefighters be raised to go to Britain to help out. The offer was accepted and the Canadian Corps of Overseas Firefighters (C.C.O.F.F.) created, recruiting experienced firefighters from across the country. A contingent of 400 men was trained in Ottawa at the old King Edward Avenue Station training tower and departed for Britain under Chief John Hough of Brantford, Ontario, in June 1942.

The Canadians arrived after the main blitz attacks were over, but saw some action in their assigned cities of Southampton, Bristol and Plymouth. The C.C.O.F.F. returned home in November 1944.

Sabotage in St. John's?

The old seaport of St. John's, Newfoundland, was also a busy link in the Battle of the Atlantic, as a naval escort base and a refuge for damaged ships. Although not to become a part of Canada until 1949, there was a strong Canadian naval presence here and, later on, an American air force base. The Canadian navy set up a well-equipped fire department to cover its various installations, occupying one of the stations of the St. John's Fire Brigade, and eventually installing a fireboat in the harbour. When the big U.S.A.F. administrative base of Fort Pepperell was built at Quidi Vidi, it had its own fire station as well.

Several fires in isolated buildings on high ground, that burned like beacons and were visible far out to sea, have been linked by persistent stories to later reports of major attacks on convoys by U-boats. Another story involves cases of arson attempts, in which the material for igniting the fire was toilet tissue.

There is always someone who cries "Sabotage!" when there is a spectacular fire in wartime, but the idea appears to have been taken seriously in the conclusions of the official inquiry into the Knights of Columbus hostel disaster in St. John's in 1942. One of Canada's most appalling life disasters — though not in Canada when it happened — the K of C fire stands out as a dramatic case of negligence and ignorance in fire safety precautions, occurring in the rush of make-do during wartime. According to Sir Brian Dunfield, who held the inquiry, the conditions existing at the time of the fire provided an ideal opportunity for someone bent on mass murder.

Many efforts were made on a charitable basis to provide entertainment, relaxation and some home comforts to the thousands of servicemen who were stationed, or in transit, far from home. Hostels like those of the Salvation Army and the Knights of Columbus were standard fixtures in or near any major base or concentration of servicemen. The buildings were often standard wooden huts, as used by the military. Accommodation, for civilians too, was hard to come by.

The Knights of Columbus established a hostel in St. John's, in a light wood-frame structure they put up on Harvey Road in the centre of the city. It was two storeys and provided dormitories, a restaurant, lounge and an auditorium dance hall. Regrettably, in the wartime rush to get things done and with weak resources, there were many things wrong with the structure. No building permit had been taken out, and this seems to have escaped notice.

Besides being highly combustible inside and out, the windows of the auditorium were covered over with plywood as a blackout measure, since St. John's was under full blackout. The three exit doors of the room opened inward and were locked. The main entrance was through a lobby and restaurant area containing tables and chairs. There was no emergency lighting.

Paper decorations for the Christmas season were draped about the ceiling. It was in many ways the same setting for disaster that existed in the Coconut Grove nightclub in Boston. That one set the stage for a fire that snuffed out the lives of 492 people, two weeks to the day before disaster struck in St. John's.

On the evening of December 12, 1942, the Battle of the Atlantic was still on, the air was cold and damp the way coastal winters are near the North Atlantic. In the warmth of the K of C hostel, the Saturday night barn dance was on. There were about 100 people in the dormitories, lobby and restaurant, and 400 crowded into the auditorium to enjoy Uncle Tim's Barn Dance program. This was being broadcast live over radio station VOCM. The crowd were Canadian, Newfoundland, British and American servicemen and their partners, the girls of St. John's.

The broadcast was on and a cowboy song being sung by a male vocalist, when over the radio waves came a scream of "Fire! Fire!" in a panicked female voice. The M.C. took the mike and appealed for calm. Muffled sounds followed, and then there was silence. The operators at the radio station and thousands of listeners had been rivetted in attention, then left in the dark.

A witness, Margaret Ryan, quoted by Rasky in his *Great Canadian Disasters*, heard the broadcast in her home right across the street. She dashed to the door to see the K of C building ablaze and hot embers already flying about, while nearby windows were breaking in the heat. Men were diving from second-storey windows with pyjamas on fire. She could hear the pounding of trapped victims inside, trying to get out. Her father, the barn dance's "Uncle Tim," and a brother escaped, but her other brother was lost. Her father had smashed the blackout screen from a backstage window, letting out a number of those on the stage.

The cry of fire heard by thousands of people in so dramatic a fashion came when flames shot out of the projection booth at the rear of the hall and booming sounds came from overhead. Fire was finding its way from confinement in the attic space above, coming down through the ceiling, escaping after building up a great heat and volume of gases. The paper decorations caught and dropped, flaming, on the heads of the crowd below, as they pushed towards the exits and the entrance lobby. Hair and clothes on fire, the screaming, panicked dancers found the exits locked, and attempts were made to break the doors down. The lights went out, people stumbled over fallen chairs and fell in the crush at the exits, piling before the doors. Two doors were broken open and many escaped, hair and clothes gone, in a state of shock.

The fire, with its deadly gases, pushed with phenomenal speed through the flimsy building. The clerk of the hostel tried to call the fire brigade by telephone at 11:11 p.m., but while waiting to get through on the overloaded phone system was overtaken by the flames and died. Some lingered to help others and lost their lives for the delay.

The ferocity of this fire came from the sudden release of pent-up superheated gases in the closed attic space across the top of the auditorium. An anonymous soldier had been seen to open the door of a storeroom, off the second-floor hallway and built into the attic space. Seeing a fire, he ran off, leaving the door open. A sailor saw this and closed the door on what he described as an intense fire in burning boxes of toilet paper. He warned his dorm mates and dashed through the rising heat waves to the lobby below. Witnesses described the flames that first intruded through the auditorium ceiling as blue. This has been surmised as being burning carbon monoxide, forced out by what firefighters call a backdraft, a sudden expansion of flaming gases as an oxygen-starved fire gets a fresh supply. In this case it was provided by the opening of the storeroom door, the fire having already penetrated the attic, turning it into an oven of hot gas.

The fire very quickly flared out of the building to illuminate the night sky. The big Central Fire Station at Fort Townshend was only about a block away across the parade square, but the firemen were oblivious to the glare, since their windows were all blacked out to keep the light in. At about 11:15, the police constable controlling traffic lights at Rawlin's Cross, a considerable distance down Harvey Road, reported to his headquarters that there was a glare of light up the hill, right in the vicinity of police headquarters. The call was relayed to the fire department at 11:17, and the first turn response was on the scene two minutes later.

Sir Brian Dunfield, who conducted the official inquiry, concluded that anyone still in the K of C building must have been dead by 11:15. The firemen were too late to save lives and were hard pressed, with all three stations turned out, supplemented by military fire companies and auxiliary firemen, to confine the fire to the hostel building. The buildings on either side were beginning to burn when they arrived.

The fire was not cooled down until the next morning, by which time little was left of the hostel but chimneys and charred mounds of debris. Then began the grisly task of digging out the charred bodies, while snow fell on the scene. The total dead numbered 99. A mass funeral was conducted for 80 servicemen and merchant seamen by the Newfoundland, Canadian and American armed forces.

Sir Brian apparently believed the fire was set in the storeroom by an enemy agent, though it was officially recorded as incendiary, possibly sabotage. A few weeks later, a lot of toilet paper was found in the Y.M.C.A. Hostel's attic. The Old Colony Club, frequented by military officers, suffered a suspicious fire soon after this. The situation roused strong suspicions, but whether it was an effort by an enemy agent or someone with a grudge against either servicemen or the hostels might be open to debate. There undoubtedly was considerable friction between local citizens and the outside forces stationed in St. John's.

Further Reductions in Working Hours

As at the end of the First World War, returning servicemen looked for a reduction in the still long duty hours in career fire departments. The major cities had been on a two-platoon, 84-hour week since 1918, but smaller ones were often on much longer hours. Firemen in Saint John, for example, were still working three days and nights, then 24 hours off, going home to meals in the old manner. Two platoons soon came in these departments after the war, and by the early fifties many departments went to 60 hours, and then to 56, which meant three shifts. However, the two working tours per 24 hours were retained, a ten-hour day shift and a 14-hour night shift, continuing the use of beds at night.

These tours remain in the eighties, though most fire departments achieved a four-shift status for a 42-hour average work week by the sixties. The shortening hours greatly increased the cost of manning, and inevitably resulted in pressures to reduce company manning. One effect was to hasten the closing of fire stations built for the horse-drawn era but no longer needed with powerful modern motor apparatus. In many cases, however, it became a redistribution of stations to new outlying districts, mushrooming in the postwar expansion reflecting the automobile age.

Firefighting Under Fire — The Bedford Magazine

The war in Europe ended in May 1945. Halifax had been overcrowded and strained by its job as the key western terminus of the Battle of the Atlantic convoy line. Long in fear of a repetition of the colossal explosion of 1917, and with a couple of close calls, the citizens and the navy could now relax a little. The VE Day riots had caused bad feelings between the city and the navy, but it was now time for a return to normalcy. Troopships were bringing the forces home, and the antisubmarine naval vessels were coming home for decommissioning. The neon lights came back on, and the merchant ships were resuming their peacetime colours. The war had not finished with Halifax and Dartmouth, though, for they had not had their explosion yet.

Before the many naval vessels could be decommissioned and laid up so the volunteer war service crews could go home, the key job was to de-ammunition them. This should have been a slow process, as safety rules allowed only one at a time to unload its lethal cargo on the jetty at the big ammunition storage magazine at Burnside, above Dartmouth on Bedford Basin. However, it was politically important to speed the servicemen home as quickly as possible, and the rules were bent despite protests. In an inordinate rush, three vessels' stocks of ammo were on the jetties at a time. The big, carefully-designed magazines in the rocky hillside on the wooded far side of the basin were very safe when the rules were observed.

The St. John's magazine had been closed and the ammunition sent to Burnside. Many ships had dumped off their stores, and the magazine were filled up. Still the ships came and the depth charges, shells, torpedoes, hedgehogs, pyrotechnics and other nasty things began to pile up in the open and even on the jetties. There was great pressure on the magazine personnel to get the ships unloaded and away, so all the rules were broken.

Trouble came at 6:30 p.m. on the hot summer's day of July 18. Apparently a pyrotechnic rocket ignited, setting off fires and a series of small explosions on the south jetty, which had a barge tied alongside. Then there was a big bang. The ground shook and the jetty and barge disappeared. A civilian patrolman died in the big blast, having given the alarm and returned to fight the fires that were breaking out in the dry grass among the piles of high-explosive devices. The Bedford Magazine Fire Department turned out. The big thunderclap reverberated across Dartmouth and Halifax, and a mushroom cloud rose high in the air across the waters of the big basin, raising fears and memories in the minds of the older generation of Haligonians.

Captain Robertson, commander of the naval dockyard, was at dinner with his wife at the Nova Scotian Hotel, a Sunday night out. He heard the boom and saw dust shaken out of the hotel air conditioning registers. With a pretty good idea of what was up, Robertson dashed to the top floor of the hotel and looked north to have his fears confirmed in the pillar of smoke. Boarding his speedboat, he hurried up the harbour and across Bedford Basin. The doughty firetug *Rouille* also sped to the scene, standing by the remaining north jetty. The *James Battle* appeared, but stood off at a safe distance.

Naval firefighters, under the guidance of the ammunition foremen, the only people who knew the layout, went to work to try to control the fires that were spreading in the dry grass and brush near the open piles of ammunition. The nearby barracks burned down; other buildings were blown flat. Keeping clear of the shells and casings they worked on the blazes constantly being set by flying ammo, risking their lives every minute.

The water main had been knocked out and its supply tank wrecked. Eventually the breaks were isolated and *Rouille* came in to the north jetty to pump water into the remainder of the hydrant system. It was difficult to tell what was going on with all manner of missiles whizzing in all directions like a battle on the Western Front. Captain Robertson hid behind a handy reinforced-concrete building, peering out to observe, but was persuaded to move when advised that it was the detonator storage building.

No sooner had he moved away than it blew up, and he was thrown into a pond. Suffering from concussion, his jacket blown off, Robertson wandered about muttering, "That bastard Mackenzie King won't give my wife a pension!"

The phones went out, but reinforcements had arrived. Although many ammunition workers volunteered, even the naval parties on the scene were kept to a minimum. The north end of Halifax was evacuated as far as North Street, most of the citizens needing no urging, but some stayed to watch the show. It was a spectacular night, full of bangs, booms and pyrotechnics.

Among several big explosions, the largest came at 4 a.m., when a concentration of over

360 depth charges and bombs went up, leaving a huge crater. The shock wave crossed the broad expanse of the basin to ricochet through the streets of downtown Halifax, breaking windows alternately from side to side. Stocks of glass, left over from supplies brought in after the VE Day riots, came in handy. This explosion was the climax. The booming, banging and whizzing went on for several days, but the worst was over.

Naval firefighters, under Captain Robertson and Carson, now a lieutenant, worked to isolate the grass and brush fires from the big magazines and open piles. The job was dangerous, with bullets flying and explosives scattered all about. Occasionally men were hit by shrapnel. A brush fire had been set at the railway siding some distance away, where several boxcar loads of explosives were set on fire and a tough battle for control was fought. Using portable pumps and drafting water from ponds, Robertson's men, their wounds patched by a courageous nurse right on the spot, achieved victory after several hectic days. It *was* a victory, considering only a third of the magazines themselves had burned.

The heroic work of the navy and its naval and civilian fire service appeared to help heal the rift between the people of Halifax and the senior service. The politically inspired and hurried de-ammunitioning of the warships, with piling in the open air, was the culprit in this episode. It could possibly have resulted in a sympathetic explosion of the main magazines and another, greater disaster than that of 1917.

The war was over and no medals were forthcoming for the courageous firemen and civilian magazine supervisors. Captain, later Commodore, Robertson was invited to give a paper at President Truman's fire prevention conference in 1947, covering his exploits in Halifax during the war. This resulted in his meeting George Richardson, long-time president of the International Association of Firefighters, and being made an honourary life member of the union.

A pier and numerous buildings were obliterated, and a wide area left blasted like a battlefield in the Bedford Magazine explosions opposite Halifax in 1945.

Large scale conflagrations still occur occasionally in modern times. It only takes the right combination of circumstances.

7

The Modern Era (1945-1980)
(XXII) Postwar Problems

Changing Cities and Towns

The physical nature of cities began to change in the construction that followed the Depression — World War II period. The automobile was a key factor in this. Truck transport, an automobile for every family, expanding road networks, and the demand for spacious suburban living caused a spreading out of the densely built community previously based on street railway lines, unpaved streets and horse-drawn wagons. Industry moved either to the periphery or outside the city limits to accommodate truck traffic or avoid taxes. To suit the automobile owner, new suburban shopping areas were built in the form of single-building shopping centres, and eventually these acquired enclosed pedestrian streets or malls. Such new departures in architecture presented new problems for the fire service and writers of building codes. However, when new, large shopping centres were totally sprinkler protected, they provided strong built-in fire protection for what was, in effect, an entire shopping district by traditional standards. This was a boon for the suburban municipality with a fire department stretched by weak finances, for the new luxury of space had caused greater costs to the municipality in longer streets to pave, longer water and sewer lines, and more fire stations.

In some cases, the old secondary ring around the traditional city shopping and business district, consisting of old industries, deteriorating housing and small businesses, decayed and became a fire-prone area with lack of upkeep. In a more recent cycle, expansion of the inner business district has been displacing the decayed area with new, taller buildings, reducing the hazard again. The modern high-rise building, detached and with parking lots alongside it, replaced the contiguous older structures, reducing the inter-building exposure hazard, but in turn it introduced new problems to be solved by firefighters and code writers, due to its height.

Eventually some industries and commuting city workers moved out so far that they were in the country and scattered so far that it became impractical to provide water mains and hydrants for them. Fortunately, this roughly coincided with the development of fog firefighting techniques and the ability to control a considerable fire with the small amounts of water that could be carried on board fire trucks. Sophisticated tactics using fast-attack pumpers, supported by large tankers shuttling back and forth from a water source, have made effective semirural protection feasible, although distances and long response times are a drawback.

Rimouski and Cabano

The dawn of a new era at the end of World War II did not mean there would be no more conflagrations. The legacy of past construction practices, the weather elements, and weak water supply systems and fire departments would still turn up in the right combinations from time to time, although less so in larger cities.

Within four days in May 1950, the Province of Quebec was visited by two conflagrations. They had in common the same principal contributing factors: severe exposure between a lumberyard and the town, high winds, dry weather, flying brands, and lack of sufficient firefighting manpower and equipment.

The first victim was the Town of Rimouski, a small port of 15,000 population on the south shore of the St. Lawrence River. On May 6, a high wind developed, reaching 60 to 75 miles per hour by 5 p.m. This snapped a pole carrying high-tension electric transmission lines into the lumberyard of Price Brothers mill. Arcing set fire to the lumber piles and the wind spread it to the whole of the yard, containing 15 million board feet of drying lumber. Two internally sprinklered mill buildings were engulfed by the roaring, windblown flames. Hot brands soared aloft on the rising wave of heat, to be carried downwind across the 200-foot-wide Rimouski River into a residential district in the main part of town.

There were not many combustible roof coverings, and this contributed in preventing complete destruction of the town, but flaming boards came sailing down among the wooden houses, even crashing through windows. Dozens of homes were set on fire. Calls for help from other fire departments were answered, but the plank deck of the bridge that was the main approach to town burned and curious motorists blocked roads from other directions.

Eventually, a number of outside pumpers appeared on the scene, only to find the town water supply too weak to supply them. The fire spread through the night, penetrating the central part of the town. The 250 patients in St. Joseph's Hospital were evacuated twice, first to a home for the aged, then to Mont Joli and other towns, as the fire burned the hospital and later threatened their refuge. The 300 aged and orphans were moved out, too.

The great bulk of the stone-walled Rimouski Cathedral was in the path of the advancing conflagration, but it helped block the fire and provided a rallying point for firefighters marshalled from a wide area. By morning the wind had abated and shifted. The firefighters finally gained control by midnight. No lives were lost. Property loss included 319 homes, 20 stores, a theatre, hotel, the courthouse and jail, the hospital, a large new technical school, a convent and seminary, totalling $16,000,000.

A few days later, on the 9th, the lumber mill town of Cabano, in the interior of the Gaspe Peninsula, some 50 miles to the south of Rimouski, lost 118 homes and 13 business establishments. Winds in this case were only a brisk 25 to 30 miles per hour, when, at 10:30 a.m., sparks from a waste burner at the wood planing mill landed in a stack of 85 cords of dry maple. This burned rapidly, the flames jumping to a nearby garage and a wooden hotel. From here, fires were set in four parts of town, and within 15 minutes more than 20 buildings were burning. The small volunteer fire department was helpless.

The conditions were similar to those in Rimouski, except that the larger lumber mill in town and its storage yard were burned later in the fire, as the flames spread across the town downwind. Six neighbouring town and village fire departments came to help, but control came only when the wind died down at night. The loss was $6,000,000 in a town of 3,200, 1,800 of whom were made homeless.

Climax of an Era — S.S. Noronic

For about a century Canadian steamships had plied the Great Lakes, providing passenger services. Railways took over the business travel and ships were left with the comfortable, relaxing cruise trade. Before the automobile and modern roads killed off the luxury travel by ship as well, it had a golden era covering the first half of the century. By the end of the Second World War, these big floating palaces were aging and slowly being retired, but they hung on as long as they were seaworthy and earned their keep. The climax of the era came on Friday, September 16, 1949, when the last of the very large luxury liners, S.S.*Noronic*, came sailing into Toronto harbour from Detroit on her last cruise of the season. On board the 362-foot, 6,900-ton vessel were 524 passengers and a crew of 171. The passengers were mostly Americans from Michigan and Ohio.

Noronic, a steel ship built in Port Arthur, Ontario, in 1913, and beautifully finished inside in elaborate wood panelling, slid into her berth at Pier 9. Across the slip was the smaller passenger cruise liner *Kingston*, while across the pier was *Cayuga*, utilized in day trips and moonlight dance cruises out of Toronto.

Many of the passengers of *Noronic* went ashore sightseeing, returning aboard to retire for the night, others partied and danced aboard. The crew all took shore leave, except for a mere corporal's guard of 15 left on watch — seven in the engineroom, four bellboys to look after any passenger needs, three deck crew and a deck officer in charge. At 1:30 a.m. the orchestra stopped, ending the dance, and the ship quietened down for the night. Half an hour later, one of the last stragglers, Captain Taylor, *Noronic's* master, returned from his evening ashore and headed for his cabin. Near the stern on the port side of C Deck, the middle one of five decks and occupied mainly by staterooms, was a linen closet used by the room maids. One of their chores was tidying up the passenger cabins in the evening. Wastebasket-type trash was deposited in a refuse container in this closet. In the quiet of that night, as most of the people in *Noronic* slept, something in that closet was burning. It was probably that superb fuse, a cigarette butt, or possibly a short circuit in an electrical panel.

Noronic was, inside at least, a beautiful and luxurious ship of a bygone era. There were thousands of board feet of painted and varnished wood in her partitions, panelling and decks. The paint was thick from many annual refurbishings. There were no fire-resistant or smoke-control bulkheads with doors to segregate parts of the ship from one another. Firefighting equipment included a number of inside and outside hoselines and numerous hand fire extinguishers, but nothing automatic. The fire alarm system consisted of a number of manually-operated pull stations that registered in the officers' quarters on the bridge deck. Procedure was that the officer of the watch would decide what action was warranted and, if necessary, sound a general alarm throughout the ship over a circuit of klaxon horns and give the fire signal on the ship's whistle. There were two "special officers" aboard who made rounds at night with a watchman's punch clock, punching in at stations on each deck, bow and stern.

The federal Department of Transport shipping inspections were very cursory, and procedures on board were slack. On fire drills the hoselines on deck were operated, but not those inside. There was a fire muster procedure on paper, but it did not contemplate conditions when the ship would be in port, only when at sea. The special officer who was to go on duty at midnight that night was not wakened as he expected and overslept, so the two wheelsmen on gangway duty did the first patrol for him, apparently taking shortcuts. This was the situation in which the small smouldering fire began to grow in the closed and locked linen room.

Returning to his cabin on C Deck from the ship's lounge just before 2:30 a.m., Don Church of Ohio, one of the last retiring passengers, saw a haze of smoke in the passageway. He traced it to the linen closet, where he heard a crackling sound coming through the locked door. Church ran for help, finding the senior bellboy, who got the door open as smoke grew thicker around them. Smoke and dull, glowing flames were visible inside. The bellboy used a fire extinguisher to no effect. Then they got out the nearest hose, and Church turned the valve. Only a trickle came from the nozzle. Considering the lack of use, the control valve had probably seized partway open.

With the open door giving it air, the fire began to intensify and billowed out of the closet into the passageway lined with sleeping cabins. Bellboy O'Neill dropped the hose and ran to sound an alarm, then went to the gangway. The wheelsman got the message and dashed up to the officers' quarters on the top deck to advise them. The officers rushed out and saw the rapidly billowing smoke coming from the decks below. The klaxon system was sounded and the ship's whistle blown, but instead of blowing the fire signal, the latter stuck to produce a confusing blast. There was a public address system throughout the *Noronic*, but no use was made of it.

The little Toronto fireboat Charles Reed in action at the S.S. Noronic holocaust, September 16, 1949.

According to Church, the fire was rolling across the surface of the varnished woodwork along the ceiling at a rate that made him feel it was already out of control. Captain Taylor, the chief officer and the second officer proceeded to run about the ship, variously banging on doors and trying to use hoses on the fire that was now racing up the stairways and through the public rooms. No one took charge, least of all the captain, according to the conclusion of the official inquiry under Mr. Justice R.L. Kellock. The few of the crew on duty also worked aimlessly to help passengers find their way ashore or to fight the already lost battle with the flames. It was too little, too late and too ineffectual.

Most of the passengers had been asleep. Those who were aroused had very little time. There was a panic of people rushing about the narrow passageways confused by the smoke, the flames rolling along the ceilings over their heads as they tried to find a stairway not on fire. Some had their clothing on fire. A dock watchman said he saw flames burst from a window located toward the stern of the *Noronic* and ran to call the fire and police departments in the office. Going back outside, he saw a mass of flame covering the after half of the side of the ship next to the dock at all deck levels. The fire had spread into B Deck, occupied mainly by the large dining saloon and public lounges. Above this, on A Deck, 94 passengers were accommodated.

The Toronto Fire Department received the telephone call at 2:38 a.m., the companies from the downtown Adelaide and Lombard Street firehalls arriving three minutes later. The little fireboat *Charles Reed* began the slow trip across the harbour from the Toronto Islands, as well. Deputy Chief Jim Stevens took the run from Adelaide Street, and while still some distance away he saw the red glare against the sky. He called for a second alarm assignment of companies by radio. On his arrival a third alarm was promptly asked for. The big luxury ship was engulfed almost completely in about a quarter of an hour.

The fire department was faced with a scene of pandemonium. The ship was an inferno for most of its length, passengers were leaping from the decks onto the dock or into the water, and the dock was filled with dazed and injured passengers, many in a state of shock. Tackling a situation new in their experience with the discipline of professionals accustomed to nasty surprises, the firefighters raised ladders up to the higher decks and lowered others down into the water to rescue terrified passengers from both directions. Aerial ladders, one wood and one steel, were extended at an awkward angle, horizontally almost, to the bows of *Noronic* to collect a number of people taking refuge there. At the extreme angle, and with many crowding on it, the old wooden spring-hoist ladder snapped, dropping some into the harbour waters.

Hosestreams seemed ineffectual on the great heat in the steel oven the ship had become, and firefighters took a beating trying to get off as many passengers as they could. S.S. *Kingston* backed out of her slip to escape the showering sparks, then *Cayuga* followed. A water taxi came by and started picking up those struggling in the water. Approaching close to the burning ship, panicking passengers jumped onto the taxi, smashing in its cabin roof. The *Charles Reed* joined in the action when she arrived. Both heroism and callousness were witnessed among the passengers and crew of the *Noronic*.

It took two hours for Fire Chief Sinclair and his forces to bring the raging fire under control, and five hours before the decks cooled enough for the firefighters to begin the grim task of overhauling the wreckage for charred bodies. By now *Noronic* had listed heavily, then settled until her stern was on the bottom. The pier was gradually cleared of its chaos of injured and bewildered passengers, as ambulances and taxis were concentrated on the scene to take them away. The major hospitals nearby mustered their staffs, the brunt of the work falling on St. Michael's.

At 7:40 a.m. firemen boarded to verify the worst fears. They found jumbles of bodies, charred and unrecognizable, an experience that sickened even those accustomed to deaths in fires. The episode is remembered vividly by some of them more than 30 years later. The final toll was 118 dead, one fifth of the passengers, but none of the crew. This fact roused public indignation.

Next day *Kingston* and *Cayuga* resumed their cruises, but the fate of the *Noronic* had a profound dampening effect on the atmosphere, to say the least. The twisted, gutted wreck was stripped, raised and towed off to be scrapped. Her sister ship *Hamonic* had caught fire from a burning pier shed at Port Edward, Ontario, July 17, 1945. In this episode, 247 passengers and about 150 crew escaped, many by sliding down ropes. Others were picked off the bow and lowered to the ground in the bucket of a clamshell coal crane. Fire spread through the wooden shed and into the ship rapidly, so it was a hurried evacuation. All remained cool, however, and only one life was lost. The ship was destroyed.

A federal inquiry was promptly held into the loss of the *Noronic*, under Justice R.L. Kellock of the Supreme Court. It was found that no one in a responsible position in connection with the ship did the right thing. No officer took charge in the emergency, and the captain acted as an ordinary sailor. He was caught with no emergency plan for a fire when in dock. The court concluded that the disaster was caused by wrongful default of the owners and master. Claims against the line cost them $2,000,000, and Captain Taylor was suspended, but he elected to retire.

The demise of the *Noronic* resulted in the installation of automatic sprinklers in most of the large inland and coastal passenger vessels in Canadian waters, along with tougher

Firefighters engage in the heart-rending task of counting and recovering 118 bodies from the gutted passenger steamer Noronic at dockside in Toronto Harbour.

rules on compartmentation and crew training. These led to the early retirement of the older passenger vessels. The *Noronic* disaster was the result of age and lack of updating of the ship, as well as ignorance and complacency with regard to fire, an utterly implacable enemy.

Winnipeg's Time Building Fire

High winds and a head start can beat the best of fire protection measures when the fuel is available. Firefighters know they are in for a tough fight with these conditions. The Time Building fire in Winnipeg on June 8, 1954, was an example of a strong fire department with a very powerful water supply being put to the supreme test: stopping a wind-driven fire in the heart of the business district. Many of the key elements in the case were very similar to ones faced by the Toronto department a quarter century later, both even involved serious exposure to a major store of the same company, T. Eaton Company Limited.

That night in Winnipeg, anyone with sense was huddled at home during the powerful windstorm from the south-southwest, ranging from 52 to 72 miles per hour. A police officer was doing his rounds, though, and after 1 a.m. discovered a fire burning behind a large neon sign on the second storey of the sprinkler-protected seven-storey Time Building on Portage Avenue. This brick, wood-interior structure was erected in 1905 and had been

Winds and winter cold has made for some memorable fires in Winnipeg, but a powerful high pressure water system was often a great ally to the firefighters.

There is the occasional exhilaration of saving a life, but there is also the frustration of knowing you did not quite make it. Winnipeg firefighters carry out a small body.

much remodelled, with resultant false ceilings and other concealed spaces. It housed two clothing stores on the ground floor and 73 tenant offices on the floors above.

The fire department was summoned at 1:18 a.m. They soon discovered that the fire was coming from within the building. Unfortunately, it had got into the old nonfirestopped wall and ceiling spaces, spreading out of reach of the automatic sprinkler heads within the rooms. The firemen also had difficulty reaching it. Not until the fire had gained considerable headway in the labyrinth of concealed spaces did the sprinklers work. But it was too late. The sprinklers were overpowered and the flames took over the two top floors, then burst through the roof.

In spite of the high winds that scattered their hosestreams, the firefighters held the flames to this building until 5:50 a.m., when the east wall collapsed into Hargrave Street. Fanned by the wind, the intense heat immediately ignited the three-storey Dismorr Block, across the 66-foot-wide street, then into an adjoining five-storey building as the fire department mustered big master streams supplied by the high-pressure fire main systems. At six o'clock the west wall of the Time building fell onto another lower one, adding it to the big bonfire.

For a time the block-long, eight-storey department store of T. Eaton Company, diagonally across the intersection with 132-foot-wide Portage Avenue, was threatened. A well-organized private fire brigade in the store extinguished smoking window frames and cornices to save this huge establishment, using 21 hoselines supplied by the company's private water system.

The Winnipeg Fire Department utilized 36 lines of 2½- and 3½-inch hose, and had 26 fire companies at the fire in answer to four alarms and extra calls. Only one was left in quarters to cover the city. The high-pressure system earned its keep that night, as it prevented a general conflagration through the business area.

Hurricane Hazel and Other Odd Jobs

Historically, firefighters have always been called upon to do more than fight fires. The fire department represents an organization of able-bodied personnel, accustomed to quick decisions and action under emergency conditions. They are also equipped, incidental to their work of firefighting, with heavy equipment suitable for gaining access to trapped and injured persons, and to provide breathing apparatus, resuscitation, emergency lighting and first aid.

From cats in trees to workers buried in cave-ins, it has been the firefighters who were called upon. These extra chores have often been dangerous and contribute to the high occupational hazards of the job. One modern activity frequently calling for the fire department's services is highway extrication following vehicular accidents. Most departments are now equipped with sophisticated tools capable of cutting an automobile apart in short order.

One memorable example of fire department response beyond the call of firefighting duties was in the floods caused by Hurricane Hazel in the Toronto area in 1954. This hurricane was rather unusual in its course into the interior of the continent, far from the usual coastal track, and it caught Toronto and suburbs unawares, lambasting the surrounding countryside with an unprecedented four inches of rain in 24 hours. This torrent made the various rivers run over their banks with savage power, causing floods in the heavily populated area. Dozens of bridges were carried away and great trees disappeared, while houses and cars floated off downstream, often with people still in them. The rivers Don, Humber, Credit and lesser creeks ran amok during the wild night of October 15 and the day following.

Much of the damage was caused on the east and west outskirts of Toronto. The volunteer fire departments of the various small suburbs, such as Long Branch, Islington and Kingsway-Lambton (later amalgamated into Etobicoke), turned out to rescue trapped citizens from the rampaging waters. They used ropes and ladders, and waded up to their necks carrying people to safety. The Dundas Street Fire Hall in Islington was pressed into use as a morgue, as victims were retrieved.

The disaster struck the fire service in its unflinching response, as well. A major calamity befell the Kingsway-Lambton department when one of its trucks, manned by volunteers, became stuck in the mud from the flood's rising waters. The five-man crew was trapped on top of the truck for three hours as the water rose. The hose floated away and then the truck itself rolled over, the crew swept away on the current. It was but one case of firefighters losing their lives trying to save others in circumstances other than at a fire.

All that was left of the Kingsway-Lambton fire truck after Hurricane Hazel, 1954.

The Cold War and Civil Defense

The fifties found Canada living in the atmosphere of the cold war and the Korean War. This brought the return of a civilian defense organization involving components at federal, provincial and municipal levels. Compared with the World War II organization, this one did not concern itself so much with a large force of manpower, such as auxiliary fire companies, as it did with organizational preparedness. Regrettably, in that period the emphasis was heavily upon preparedness for nuclear war, as opposed to a more general organization that would be useful in all manner of disaster situations, as is the current concept.

The federal government took the lead, and its jurisdiction was in the area of war-related disasters. The result was a lack of everyday practical application, in which

Civil defense fire resources were reestablished during the Cold War in the 1950s, but civilians could not be trained to extinguish nuclear fires as they were World War II incendiaries in this photo.
- PAC-PA108263

Fire departments have long engaged in a variety of emergency work other than firefighting, notably in the areas of extrication and rescuscitation. A large rescue squad truck of the 1960s holds a great variety of specialized equipment for such purposes.

Pioneering application of water-fog techniques on a major fire in Hamilton, 1955.

professionals like the police and firefighters had a real interest. A Civil Defense College was established at Arnprior, Ontario, where many municipal fire officers took rescue training courses of value, and matching federal and provincial funds were made available to help strengthen municipal fire forces. Ontario made particularly good use of this, standardizing all the hose and hydrant threads in the province and setting up permanent county-wide mutual aid fire defense organizations. The federal government also distributed 50 small and medium-sized fire department pumpers about the country for training purposes. These proved useful in major fires when fire departments were hard pressed.

When the cold war atmosphere abated, civil defense was whittled down by government budget knives, its greatest deficiency as a useful organization being its lack of a close link to the commonly occurring civil disasters, or a recognition that fire, police and other ongoing emergency professionals were the real experts and those who would be doing the real work in any civil or wartime disaster. Thus much-needed disaster organizations fell on thin times, until their comparatively recent rejuvenation under the Emergency Measures Organization programmes of most provinces.

Severe disaster potential in the transportation of large quantities of dangerous chemicals and flammables has encouraged this new thinking. A salient example was the disbanding of the Ontario Emergency Measures Organization several years before the famous Mississauga train wreck emergency of 1979. Luckily, the City of Mississauga had well-developed plans of its own. The province has since reestablished its organization, but with a shift of responsibility for organization toward the individual municipality.

A little-publicized aspect of fire protection during the Korean War period, as during the Second World War, was the free advisory and inspection assistance provided to the federal government by the fire protection engineering staff of the Dominion Board of Insurance Underwriters and the Canadian Underwriters Association. Their specialists tested water main supplies, over a thousand trailer pumpers, checked sprinkler and other fire system plans, and assessed the protection conditions in bases of the armed forces and vital industries. This ended with the setting up of equivalent professional staff within the government in the 1950's.

Modern Fire and Building Codes

From earliest times, building bylaws, which dealt largely with fire considerations, and fire prevention laws were a matter of local municipal enactment and were locally written. After the middle of the nineteenth century, there came model building and fire prevention codes prepared by the National Board of Fire Underwriters in the United States. These codes were prepared by engineers in consultation with fire chiefs and building experts and recommended to cities. The latter adopted them or adapted them as they saw fit. More recently, in the 1950's, the Dominion Board of Insurance Underwriters provided a Canadian set of codes, similar to the American ones. Organizations of code-enforcing officials in the U.S., and the National Fire Protection Association in Boston, also entered the field.

With a stronger provincial involvement in fire prevention enforcement following World War II, supplementary code development to these traditional sources was sought. In response, the federal National Research Council set up a fire research section in its Building Research Division in 1951. Some work had been done previously in the writing of a National Building Code of Canada, to serve as a model code for municipalities to enact if they wished. A programme of volunteer committees made up of representative experts, backed by the council's research facilities now began the system of regularly revised documents, the National Building Code and the National Fire Code, and these became widely adopted by cities and towns.

The insurance industry dropped the promotion of model codes to become part of the national committee system of broad-concensus code writing, as many involved interests pitched in to try to keep the codes for public safety abreast of fast-changing architectural concepts, including enclosed mall shopping centres, atrium buildings, tall buildings, and underground cities, all posing new problems in fire protection.

A shift from the local responsibility for enactment of these codes came during the 1970's as the provinces began to adopt the national codes as provincial codes, superseding the municipal codes. In some cases, however, the provinces have made modifications in the model codes, detracting somewhat from their countrywide uniformity, which would be a boon for architects and builders.

An important part of the enforcement of codes and standards in fire protection is the provision of approvals and testing of materials, equipment and components. It was pioneered by the insurance industry through establishment of Underwriters Laboratories Inc. in the United States, and provided its label or listing on a great variety of items it concluded were safe according to its professional standards. In Canada, the similar Underwriters Laboratories of Canada was set up in 1920, and engaged heavily in the approval of electrical equipment. This work was soon passed on to the Canadian Standards Association. After a hiatus, U.L.C. was reestablished following the Second World War, and today these two nonprofit organizations provide a wide range of testing and standards-writing services.

Fire testing of building components, like this firedoor about to go into the furnace at Underwriters Laboratories of Canada in Scarborough, was started by the insurance industry and has become part of the modern building code system.

XXIII Recent Challenges

The Volunteer, Still Alive and Well

Volunteer firemen disappeared from Canada's major cities by the turn of the century, but in large and small towns, villages and rural areas, they are still alive and, for the most part, well. There are about 70,000 of them, serving isolated villages like Mayo, in the interior wilds of the Yukon, or sharing the load with a career force in a city of over 100,000 population like Burlington, in the Golden Horseshoe of Ontario.

The enthusiasm of the volunteer is as keen as it ever was in some parts of the country, less so in others. Possibly the most notable hotbed of volunteer firemen is Nova Scotia, where some 7,000 of them raise much of the funding to purchase modern fire apparatus and equipment, and to build their own fire stations. They were also instrumental in the establishment of their own province-wide training programme.

Across the country and to the shores of the Arctic, firemen turn out in evening and weekend training sessions, helped by provincial mobile instructors, taking pride in their vital community service. The stereotype of the volunteer fireman as an ineffectual amateur is seldom accurate today. Many of them, in fact, object to the career firefighter adopting the word "professional" as his exclusively, for the volunteers are often very professional themselves, in the formal sense of the word.

Volunteer and composite (part-career, part-volunteer) fire departments today protect high-rise buildings, large shopping malls, and major industries with aplomb. Modern radio communications networks, mutual aid plans and an increasing density of neighbouring communities within reasonable response times have helped bring strong firefighting forces to bear when necessary in the small-town and rural setting.

The volunteer fire company still functions as a sort of service club in much of the country, providing an indispensible public service, a social side, often other fundraising causes and charitable enterprises. Like their career counterparts, the volunteers have a sense of brotherhood made close by the sharing of vicissitudes and achievements. The person we call a volunteer fireman may be a true volunteer, in the sense of working entirely for nothing, but most receive a small stipend, enough to pay their dry cleaning bills and out-of-pocket expenses. A number are paid real wages on a per-call or per-hour basis — what used to be called "call firemen."

Fundraising projects are an important element in the work of ambitious fire departments, because they want to do a good job with the best equipment. Some of the funds go for paneling their clubrooms, perhaps, but first priority is for firefighting and lifesaving gear. The Annapolis Valley towns of Nova Scotia, running from Windsor through Kentville and Wolfville to Yarmouth, are famous for their state of the art apparatus and equipment sitting in their fine stations. Upstairs lies the explanation, a well-appointed hall where dances, dinners and bingo support the enterprise that produces this source of pride and service.

Women are very much a part of the picture in the best-organized volunteer fire departments, supporting the social and fundraising activity. Active firefighting is still almost entirely a man's preserve, though there are women volunteer firemen and have been women fire chiefs. In some places, it has been necessary for the women to take over the whole firefighting job during the day, when the men are all working out of town.

The volunteers still generally elect their chiefs, and normally have the wisdom to choose someone experienced and able to command them effectively in action. The selection is, like all key actions of a municipal fire department, subject to approval of the municipal council.

The relatively isolated and self-contained town or village seems to have more cohesive community spirit than the small satellite lying close to a metropolitan area. In the latter, there is often a problem recruiting volunteer firemen and in maintaining an active interest.

Today's volunteer fire departments are highly organized and train seriously. This composite crew from several departments demonstrates a rural tanker evolution at the Atlantic Firefighters Workshop.

In addition, the men may commute to the city to work, depleting the available force during the day. In a few large bedroom suburbs this has necessitated maintenance of a career firefighting crew working the day shift only, Monday to Friday.

Volunteerism in all fields now suffers competition for people's time. In fire service this threatens a multi-million-dollar asset and is a worry to some municipal officials. Firefighting is a labour-intensive activity, demanding numerous bodies and skills in a sometimes dangerous job. Other than by the preparedness of those willing and able to accept the danger, discomfort and inconvenience for the satisfaction and fraternity of this unique service, many communities could not protect themselves. In this, it is as it always has been since the days of our pioneer forefathers.

Life in the Firehall

Whether it is the local custom to call it a firehall, fire station or *poste de pompiers*, whether they call themselves firemen, in the time-honoured tradition, or firefighters, as the modern career-type usually prefer, the people who inhabit this facility are special. They have their own unique fraternity, one typical of those who go into danger together, who suffer trials, tribulations, fatigue, discomfort and triumphs together in a common and humanitarian calling.

In the prosperous postwar decades, along with other organized labour, career firefighters obtained shorter hours and substantially better salaries. Work hours went down to 60, 56, 48, and then 42 hours a week. Tours of duty generally remained the same, a ten-hour day

Primitive drilltowers have given way to comprehensive training plants, many in larger fire departments, others operated by provinces or cooperatively. This is part of the Montreal Fire Department school, one of the first.

shift and a 14-hour night shift, with the traditional use of beds on the night shifts. Sleeping on night duty is a privilege descended from continuous live-in duty, and is not considered to delay turn out for alarms. Some prefer a 24-hour shift for its longer off-duty periods. Firefighters cook their own meals, though a fine dinner may dry up in the oven while they fight a fire.

While there is considerable free time in a standby mode, the need for instant switch to vigorous physical and mental activity makes for a stressful work situation. For this reason, exercise and physical fitness have been actively promoted in recent years. During the day and evening, in well-run fire departments, the firefighter is kept busy with station and equipment maintenance, drill and training and, since modern radio communications, fire prevention and familiarization inspections throughout their districts.

The old picture of continuous cribbage or checker games is long out of date. Television, hobbies, study courses and a variety of charitable and other projects occupy any free time in the station. For many years it was a widespread custom to collect and repair toys for distribution to needy families at Christmas. Shorter work hours and general prosperity ended this activity.

The modern firefighter is businesslike in union contract negotiations, but has always exhibited a strong sense of mission and pride in his service and uniform. Despite jokes about excessive use of axes and water, he has enjoyed a favourable image with the public.

His most common joke about himself is a self-deprecatory reference to failure to prevent destruction of a building by fire as "building a parking lot."

In spite of the very real life hazards, as one of the most dangerous of occupations, firefighting is a good job, satisfying in its severe physical and mental demands and its sense of service. There are waiting lists to join and turnover is very low. The range of emergency responses, meanwhile, expands. Some fire departments now record as little as 20 percent of their responses as fires.

The job is demanding more and more skills and technical competence, with a need for more education and specialists. Some firefighters are eager for this, others would prefer to stay purely physical "smoke-eaters," the work that originally attracted them to the field. In the eyes of municipal officials and progressive fire chiefs, the rising cost of maintaining career firefighters round the clock now necessitates their still more active involvement in the new initiatives in fire prevention. Fighting fires more by preventing their occurrence should not only change the nature of the job, but make it safer, too.

Another St. Urbain Street Tragedy

Back in 1877, when the life of a career firefighter was tough, with primitive tools, no protective clothing and the continuous-duty system, hardship produced a strong bond between the firefighters and esprit de corps was what held them together. Never had it shown more strongly than at the Oil Cabinet and Novelty Works fire on St. Urbain Street, April 29, 1877, in Montreal. When part of a wall from the blazing building fell on a crew of firefighters in an alley, there was no hesitation on the part of their buddies, who dashed in to dig among the hot bricks in a rescue attempt. The rest of the wall then fell on the rescuers. Still more men immediately went into the heat, smoke and burning rubble to try again.

That was selfless devotion at its highest peak. One might conjecture as to whether the same spirit would hold under the modern, more impersonal multiple-shift organization of over 2,000 firefighters. The answer came on the same street 86 years later. On April 6, 1963, a fire was gutting a large four-storey furniture store on Ste Catherine Street at St. Urbain. Firefighters working in smoke-obscured St. Urbain Street heard a resounding crash within the building as the upper floors gave way. They backed away from the old stone wall in apprehension, not sure what was happening. Then the cry went up, "There are men in there!" For the firefighters of Montreal, standing in St. Urbain Street in 1963, there was no hesitation either. In a minute they had ladders to the windows and were climbing into the shaky tangle of wreckage in the search for trapped men. The challenge had been met, although for three of their mates it would be too late.

New Kinds of Fires

The postwar expansion, prosperity, new technology and architecture brought with them new fire situations to challenge the resourcefulness of the fire service, the building and fire prevention code writers, and the whole technology of fire protection. In cities growing with a new population boom, people and their commerce were being accommodated in buildings of increasing heights and areas. The high-rise office, hotel and apartment building arrived and proliferated. City people also began moving underground for rapid transportation, all-weather shopping and easy movement from one building to another. Even their automobiles were stored underground.

Arrays of new processes, chemicals, bulk transport of high-energy fuels, and other problem materials with disaster potential came to perplex the fire chief and his staff. On top of all this, a significant loosening of traditional restraints in social codes was leading to lax attitudes toward the common welfare in such areas as fire safety. The right to free self-

Spectacular and dangerous fires occur in petroleum facilities and where the fast proliferating dangerous chemicals are handled. Firefighters are at a disadvantage when there is inadequate built-in protection.

expression seemed to take precedence over the security and safety of society in the eyes of a large younger generation. Growing sophistication and competence in the professional fire service of larger cities was matched by more difficult fire problems.

Problems Underground

The subway, for mass transit of large populations, came to Canadian cities in 1954, when the first line was opened in Toronto. Montreal built its first lines in time for Expo 67. The first subway fire also came in Toronto. New materials, whose fire dangers were not yet impressed upon the world, were to show what damage they could unleash.

It was on March 27, 1963, that Toronto's relatively new, clean, efficient subway ran into trouble with fire. A six-car train was proceeding south on the Yonge Street line when a short circuit was indicated in the fourth car and its power was cut off. At the next station, smoke was observed coming from under this car, from an overheated resistor. The train was ordered to proceed on to the Union Station storage track, where the fire could be gotten at, a distance of 1.9 miles. On arrival, the train crew used portable extinguishers without effect on the now growing fire. They were soon driven back by smoke, and ran to the nearest station to give the alarm. Despite the outward appearance of having little to burn in it, the subway train provided ample fuel for a hot, smokey fire in the confines of the tunnel, spreading from car to car. Each car had plastic panels and wood furring in the ceiling, cork, wood and rubber tile in the floor and, worst of all, highly flammable foam-rubber-padded seats.

The Toronto Fire Department responded with a total of three ladder companies, seven pumper companies and a rescue squad, utilizing nine hosestreams stretched in from both directions, Union Station and St. Andrews Station. The thick black smoke and high heat drove them back and made their efforts ineffectual. The whole train was destroyed, along with signalling equipment and considerable heat spalling of the concrete tunnel. Fortunately, there was no loss of life. The fire resulted in a thorough reevaluation of combustibles in the construction of Toronto subway cars and an upgrading of fire protection measures. Even so, there would be another serious fire in a subway train at Christie Street Station, again demonstrating the difficulty in controlling fires in confined underground areas.

Montreal's rubber-tired "Metro" subway trains, patterned after the original of this type in Paris, began running in the euphoria of the great Canadian centennial celebration, Expo 67. It was one of the many ultramodern technological achievements with which Canada surprised the world at this magnificent world's fair. The Montreal Metro, for all its sleek beauty, had serious flaws in the area of fire protection.

Many automated safety features were built into the system, but a decision had been made not to have water lines down in the stations or tunnel, where leakage could be a problem. Even more than their Toronto counterparts, the Metro trains held a not readily noticed volume of high-energy release combustibles. Each car had 16 rubber tires, plywood floors, foam seats, plastic interior components and 84 gallons of lubricating oils.

Like the Toronto Fire Department at their 1963 fire, the Montreal Fire Department was equipped only with the universal 30-minute duration compressed-air breathing apparatus. This was not enough time for a firefighter to work his way down a long tunnel, do some useful work and return. A new potential problem had arrived on the scene, but as so often is the case, the potential had been neither foreseen nor planned for.

The costly results for Montreal came on December 9, 1971. A late-evening train of nine cars pulled out of the most northerly Henri Bourassa Station, fortunately without passengers, as it was executing its turnaround in the storage area beyond the station. The train accelerated rather rapidly in the view of the tail-end guard, and he started to go forward through the train, but was thrown to the floor as the train suddenly stopped dead, having smashed into the end of a parked train in the storage area ahead of it.

Forest fire. - Canapress Photo Service

Montreal subway fire, December 1971. – LaPresse, Montreal

The guard and an inspector, who had been on the station platform, raced to the head of the train to find the operator pinned in the demolished front-end cab. Smoke was now pouring from a car further back. Unable to extricate the operator, in immediate danger of being trapped in the tunnel themselves, the two abandoned the victim to his fate, caused by an apparent driving error on his part. The collision had caused one car's guide wheels to jump the guard rail, and the body of the car set up a continuous electric arc with the live third rail that did not trip the circuit breaker.

The fire fed ravenously on rubber tires, lubricating oil, plastics and plywood, filling the dead-end tunnel with high heat and thick black smoke. The first-due fire companies from Station 35, just around the corner from Henri Bourassa Station, faced this impenetrable atmosphere and called for help. There were two trains parked in the storage area in the tunnel, side by side, making 36 cars in all in the ensuing incineration. Twenty-four cars were destroyed and 12 heavily damaged, at a cost of six to seven million dollars. The tunnel's reinforced-concrete roof was severely spalled.

Director Rene Plaisance brought 24 companies to the scene, many simply to provide relief crews, as there was little firefighting possible. Limited to 30-minute air supplies, and with a 1,500 to 2,000-foot distance from the above-ground hydrants to the fire, no results were being achieved. The firefighters were held 200 feet from the burning trains. Eventually the fire department decided to flood the entire double-tracked dead-end tunnel, which was on a down slope. This was done with hoselines from a ventilator four blocks away, filling the end of the tunnel to a depth of 13 feet.

Lessons were learned the hard way once again and improvements were made in fire protection facilities and pre-fire planning. In spite of this, another fire, destroying still more rolling stock, occurred later.

Other fires have occurred in the extensive below-street-level shopping areas that have proliferated in downtown districts. These have been difficult to ventilate and get at for firefighting. In less than a decade, mid-sixties to seventies, Montreal acquired ten miles of underground or windowless shopping areas and pedestrian ways. Toronto has since matched this.

The Megastructures

Large complexes and huge single buildings have been chosen in many cases to replace groups of smaller buildings for convenience and efficiency. These range from the large shopping centre, with its enclosed pedestrian malls, to the huge automated storage warehouse. It is a good thing that the enormous values involved encourage the provision of automatic sprinkler protection.

There have been many shopping centres built with, typically, a large "anchor" chain department store or supermarket at each end, sprinklered by policy of these big companies, but with a large number of small shops between not so equipped. Severe fire damage has occurred in such shopping centres across the country. One was attacked with a bulldozer and waterbombers when out of control by a weak fire department. In some cases a fire confined to one or two small shops has caused major damage from spreading smoke through the mall into numerous other tenancies.

High in the Sky

In 1950 there were fewer than ten buildings of more than 20 storeys in Canada. By the 1960s they were springing up like weeds, along with many more in the 10 to 20 storey range. Business offices, hotels and apartment complexes were moving the city upward in a new phase in city development. Until this got under way, fire incidence was low because there were few buildings of great height. However, even in New York, the home of the

The spread of highrise buildings in recent decades was well under way before the fire weaknesses in their design were established and guarded against.

skyscraper, where there were enough tall buildings to develop some fire experience, major difficulties to come had not yet arisen.

This was because the design of the tall building was changing. The Empire State Building, long the world's tallest, had been struck by an air force bomber that deluged the interior of an upper floor with flaming gasoline. Nevertheless, the Fire Department of New York controlled the ensuing fire without too much difficulty. The increase in high-rise buildings brought fires in Canada, together with a realization that something was not right. The Empire State Building and its contemporaries, like the Sun Life Building and the Bell Telephone Building in Montreal, and Toronto's Bank of Commerce, were built with brick and masonry exterior walls having windows punched in them. Even interior partitions were usually clay tile and the offices of the day were broken up into fairly small individual areas. The building codes called for several hours of fire resistance in the floors to prevent upward spread of fire, while the traditional compartmentation and solid outside walls complemented this feature.

Postwar architecture brought widespread use of the light exterior curtain wall, large expanses of glass, sheet metal and precast material hung on the building frame. Along with this there was a trend to large, open office areas, and lightweight partitions were introduced. The whole building was sealed and mechanically ventilated. Then came the use of plastics in building insulations, decor and furnishings on a large scale.

The result was fires up beyond reach from aerial ladders on the outside, larger fires because of fewer and weaker partitions, and the development of large volumes of dangerous smoke that could not easily be vented. The population of large office buildings reached the thousands. A simple fire alarm that ordered everyone to get out immediately was not practical. It would take 30 minutes to an hour to evacuate these buildings.

The new smoke was the biggest problem, leaking around fire-resisting floors by the curtain walls, spread by recirculating air conditioning systems and by the thermal updraft in stair, elevator and service shafts. The necessity of controlling the spread of smoke became a controversial and essential element in revision of building codes everywhere.

The first formal suggestion of these problems came following a minor fire in a service shaft in CIL House, a new 36-storey office building in Montreal, in December 1963. Problems due to spread of smoke, slow evacuation of 3,000 occupants, loss of lighting in exit stairs, inadequacy of traditional alarms and difficulties for firefighters were postulated. A second fire, more serious, occurred in this building, but luckily at night. The building was subsequently fitted with the first emergency public address system for controlled or selective evacuation, and was utilized by the National Research Council in studies of air movement in tall buildings, this leading to radically new building code requirements.

Other incidents, where fires occurred above the reach of firefighting, ventilation and outside rescue, joined the record in Toronto and Montreal, as well as in the United States. Luckily, loss of life, when it occurred, was very small, until several catastrophic losses occurred in very substandard high-rise buildings in Brazil and Korea in 1972.

A reminder of new fire problems needing solutions, as a result of new architecture and materials involved in tall buildings, came in a frightening experience for 200 guests at a year-old luxury hotel in Montreal in 1967. The owners of this fine hostelry, forming part of a shopping and office group, believed they had done their best to ensure safety from fire, surpassing the requirements of the Montreal Building Code. However, things that became obvious in retrospect were not considered even by professionals in fire protection.

The great danger in this case was the high volume of fuel in furniture, decorations and interior finishes of restaurants and shops in the windowless ground floor of the complex. The hotel tower rose 38 storeys high on this base. A hot unventilated fire in the restaurant area — that the fire department could not get at — produced large volumes of the heavy, greasy smoke now all too familiar, filling the ground floor, hotel lobby, elevator and stair areas. Fire departments, code writers and architects had a need to rethink their methods.

Not only real fires, but novels and motion pictures that capitalized on a growing fire awareness by the public, seemed to contribute something to a realization by architects and builders that they had a responsibility to contribute to improved fire safety in buildings, especially tall ones.

The Dominion Fire Commissioner's Office, responsible for fire safety of government employees in numerous high-rise buildings, took action to investigate the problem of slow building evacuation under traditional rules, and with the help of the National Research Council, gave Canada and its model building code a lead and set an example for the world. Entering the eighties, the public fire protection system, still heavily oriented to fighting fires more than preventing them, is struggling with the need to make existing buildings safer and new buildings fully safe from severe fires.

Hazardous Materials Spectaculars

As if changing buildings and their contents were not enough, in the 1970s there was increasing incidence of fires, leaks and spills involving hazardous chemicals in a bewildering array, occurring not in buildings alone, but also during transportation. As always, it was the fire department that was called out when there was a problem that no one else could or would handle.

As the result of some highway and railway accidents involving dangerous and environmentally destructive chemicals, in addition to highly flammable fuels, government and industry interests have joined with the fire service in planning for control of potential disasters.

One of the more spectacular transportation accidents involves liquefied petroleum gases. These gases include common propane and butane and are transported in pressurized

New chemicals and other exotic hazard sources proliferate to complicate firefighting and control of spills, frequently on highways. Firefighters practice on an old tank truck with fog streams.

tanks in liquid form. When released to the atmosphere, they in effect boil rapidly to a vapour which is highly flammable and explosive. In fire conditions these tanks may explode, and the spectacular fireball is referred to as a BLEVE, for boiling liquid, expanding vapour explosion. In the common cylindrical railway, highway or stationary tank, the ends tend to be the weakest point. When a tank explodes, one end is likely to give way and the released burning vapour makes the tank take off like a rocket, sailing for hundreds of feet.

One or more BLEVEs occurred in the famous Mississauga train wreck on the night of November 10, 1979. A train of 106 cars was running through the north side of the city toward Toronto on the C.P.R. main line. The train consist included a variety of touchy chemicals — tank cars of butane, propane, toluene, styrene and others. In the middle, for good measure, was a car holding 90 tons of liquefied chlorine. Going at 80 k.p.h. one car's wheel-truck suddenly broke up and went sailing off into the air, glowing red hot, the result of a hot box, an overheated bearing.

The pileup of 25 derailed tank cars created a jumble around the crossing at Mavis Road, in a north-side light industrial district. The train crew managed to close the air valve at the break in the train and move away with the 27 cars on the rails and still attached to the locomotives. The pile of damaged cars inevitably leaked and burst into flames, followed by a great blast as one car took off, soaring 2,300 feet through the neighbourhood, lighting up the sky in a classic BLEVE. Four more explosions followed. With luck and not too many major explosions, the fires could still be confined to a relatively small area of damage and danger, but for one problem. In the middle of the wreckage was the tank with 90 tons of chlorine, a deadly poisonous gas, and the tank had a major leak in it.

The Mississauga Fire Department, under Fire Chief Gordon Bentley, marshalled its resources and adopted the customary controlled-burning tactic. The Etobicoke fire department filled in at Mississauga stations. Because of the chlorine, breathing apparatus and spare air bottles were borrowed from various neighbours.

To extinguish a fire from a leaking LP gas tank would allow the liberation of explosive gas. The controlled burn lets it burn as it escapes, preventing an explosion. In this they were largely successful, using a number of heavy streams from monitor nozzles to cool the tanks.

The chlorine tank, with a perceived potential to spread death over a large part of the suburban city of 300,000 population, made necessary the evacuation of a considerable area. The provincial government had disbanded its Emergency Measures Organization, leaving only a committee of cabinet to handle any civil emergency. Fortunately, the Mississauga Fire Department and the Peel Region Police had a local disaster plan to bring forward, having been involved in aircraft emergencies around Toronto International Airport, located in Mississauga.

Evacuation of citizens was ordered over increasingly large areas as time went on. For the most part, the middle-class bedroom suburb evacuated itself. With a car in virtually every home and the emergency occurring over a holiday weekend, it went fairly easily. By 12:30 a.m. on the 12th, virtually all of Mississauga and part of Oakville were emptied, an exceptional evolution.

There was a great demand for breathing apparatus for both firefighters and police in the danger areas. For five days the firemen worked with police and chemical industry specialists, with breathing apparatus always at hand, despite several violent explosions. Eventually, the chlorine tank was approached and secured. Incredibly, no lives were lost, though eight firefighters suffered chlorine injuries.

This incident, involving the orderly evacuation for some days of almost the whole population of a large city, attracted worldwide attention and in Canada brought the elements of private industry and various government agencies together in emergency planning. It also demonstrated the need for emergency measures organizations, in being organized for real emergencies in great variety, happening any time, and demanding a great variety of expertise.

Mississauga train disaster, November 1979. – Globe & Mail, Toronto

Opemiska Club fire, Chapais Quebec, December 1979. - Canapress Photo Service

Increased Arson Fires

Another challenge to the fire service fraternity in Canada, in the unsettled social milieu of the sixties and seventies, was the growing incidence of deliberate fire-setting for fraud, revenge and sheer vandalism. From 1967 to 1977 these incendiary fires increased by five times, at the end of the decade reaching 30 percent of the total fire loss.

In the United States, during the 1960s, there was an unprecedented arson rampage in the ghettoes and slums, with their deteriorating building stock. The loss of community pride and a sense of frustration and outrage on the part of disadvantaged racial minorities in the major cities are considered to have been the dominant element in this. No close parallel existed in Canada, but violence as a means of social protest and vandalism as an outlet for dissatisfaction were in evidence generally. It was an era of rising crime, permissiveness, and evidence that real and fancied rights of the individual were taking precedence over the rights of the community.

This condition was the signal for vigorous cooperative efforts between the fire service, provincial fire marshals and commissioners, police, insurance investigators, the courts and

Shocking losses in lives and property occur as a result of greed, revenge and vandalism. Fire authorities fight a constant war to stop them.

other elements of the community in a number of hard-hit cities. Arson to defraud insurance companies or cover up burglary or other crimes risks lives, at very least those of firefighters. At its worst it becomes mass murder.

A shocking example of the latter was the tragedy of the Bluebird Cafe in Montreal in September 1972. Several disgruntled patrons, not satisfied with the seating offered, cold-bloodedly spread gasoline on the stairway to the second-floor nightclub, where some 200 patrons were located, and set fire to it. The resulting flash fire filled the stairway and spread swiftly into the lounges above. The only exits left were a narrow way through the kitchen to a fire escape and a locked door to another stairway not intended as an exit. The windows had steel sashes. Firefighters, responding to three alarms transmitted in six minutes, worked frantically to rescue the trapped occupants. They had little chance for success, 37 people dying and 72 suffering injuries, including 15 firefighters, a measure of their desperate efforts.

Less coldblooded, more a case of ignorance and thoughtless vandalism without understanding of the potential consequence, was the disaster in the Opemiska Club in the remote mining town of Chapais, Quebec, December 31, 1979. A patron at a Lion's Club New Year's Eve party in the wooden hall flicked a cigarette lighter at a decorative archway of fir branches set around the inside of the entrance to the hall. The more than three-weeks-old, dried-out branches flared up in flame, setting fire to the combustible fibreboard ceiling of the hall.

In some ways it was a replay of the Knights of Columbus fire in St. John's in 1943. There was confusion, a delayed reaction, the lights went out, and there was a stampede by most of the 337 occupants toward the two rear exits. Fire flashed across the interior finish, and those who escaped stumbled out into the extreme cold, hair and clothing on fire. Forty people perished inside and seven more died after getting out. As in the case of the Bluebird Cafe, there was a disastrous combination of a set fire and an inadequate basic fire safety condition in a place of public assembly.

The Blitz in Montreal

The 1960s and 70s were times when considerable social upheaval, caused conflagrations on the social, political and labour scene. Montreal, an old city with old fire-prone districts, is a tough city in which to be a firefighter, as Canadian cities go. As the seventies began, a deteriorating relationship between an increasingly militant firefighters' union and a tough, intransigent civic management resulted in strikes by the firefighters, a rarity in the Canadian fire service.

At first the firefighters stood by at the training centre, available for serious fires. The worst strike occurred in November 1974, when they withdrew their services totally. The management-level chiefs, about 35 of them with the rank of district chief and higher, tried to maintain some degree of fire coverage, as they had done before. However, this time, a number of serious fires broke out, culminating in a series of major group fires on the night of Friday, November 1.

The red glow of several big fires burning in several parts of the city, running through the congested old east-central residential blocks was reminiscent of the London Blitz fires in World War II. While the chiefs and a few volunteer helpers were fighting spreading fires on Papineau Street at Mount Royal Avenue and Wolfe Street near Sherbrooke, Eric Sprenger, an authority on old fire apparatus, turned out to offer his help at the second fire. On his way there he found still another fire had raged through half of a block on Alexandre Deseve Street and was fanning out toward Plessis Street without so much as a garden hose turned on it. The citizens were standing there, stunned, not knowing what to do.

With the blessing of his friend, the assistant chief mechanic, Sprenger obtained a pumper from No. 8 Station and, with the enthusiastic help of numerous teenagers, put a number of hoselines into action. The youthful crew, without protection of helmets, coats or boots, stuck to their hoselines through the night, soaked to the skin, aided for a time by a crew brought in from the suburban Verdun Fire Department.

As dawn broke, the pickup crews had won the day on Plessis Street but had to work on till noon, when relieved by another suburban crew from LaSalle. In the meantime, about 4 a.m., Sprenger saw yet another baleful red glow in the west, on Robin Street, then another to the south at Champlain and Maisonneuve streets. The Verdun crew was withdrawn to go to the latter location.

Teenagers fought the Robin Street fire, too, and others did yeoman service on Wolfe Street. Again, on Sunday, teenagers manned lines at another fire. An older man, able to drive a truck, fetched a pumper from a station for Eric Sprenger to operate. A mechanical engineer who had witnessed the havoc of the London Blitz, Sprenger said it was the worst fire situation in Montreal since the conflagrations of 1852.

This was both a warm story of volunteerism in the traditional mold, and a sad story of the failure of a system, failure resulting from the inability of civic management and its fire service to work together in the interest of the citizens they should both have been committed to serve.

The modern breathing apparatus that protects the firefighter in a hazardous atmosphere has evolved over a long period. For many years, dependence was placed upon the filter masks these men wear.

Firefighter Safety — Breathing Apparatus

The mid-nineteenth century city volunteers provided their own firefighting jackets and leather helmets. The first professional firefighters were issued little in the way of protective clothing for their firefighting duties and were expected to have the capacity to eat smoke or develop leather lungs. The leather helmet, as perfected over the latter half of the century, was highly successful in protecting the head from falling bricks and ceilings, but the most dangerous enemy was the atmosphere the fireman had to "eat."

During the first half of the 20th century, the leather helmet continued to provide practical head protection, and some firefighters gave it up reluctantly even after 1975. Rubber boots and a heavy, rubber-lined duck coat with snap buckles became practical and universal in Canada's winters. Their black colour was considered suitable for the dirty work of the job, until the safety feature of visibility was accepted as being more important. The old standard of chiefs wearing white and the troops wearing black began to disappear.

Active research and concern for firefighter safety has been a preoccupation in the field as firefighters become exposed increasingly to dangerous atmospheres, ones which did not occur 20 years before. The widespread use of plastics in buildings and their furnishings has been a major contributor to this problem, affecting firefighter and building occupant alike. No longer is breathing apparatus something provided for a rescue team only. In many fire departments every firefighter entering or working near a burning building is supplied with this equipment.

The supposedly leather-lunged old-timer who scorned such protection has gone, replaced by firefighters with an awareness of the hazards of their profession and knowledge that there are means to reduce them, firefighters aware of the importance of safety, competence and physical fitness to their performance.

There is a long way to go in bringing modern technology fully to bear on the question of safety to both the firefighter and the citizen in the fire situation. However, a new awareness, together with occupational safety and public safety codes, is spurring activity in this direction. The weakness in national coordination of effort in the fire field has not helped.

The National Picture — Today and Tomorrow

With protection of the public against fire a provincial responsibility, but carried out largely at the municipal level, there has been a local orientation and lack of a nationwide coordinated attack on the problem of fire destruction. Canada's fire loss record is among the worst of highly developed countries in the world, and in life loss runs neck and neck for title of worst of all with the United States.

These were among the conclusions of a study of the whole system for control of fire made by the National Research Council and published in 1980, the first such effort since Grove Smith's report for the Commission on Conservation in 1918. An effort to meet the challenge of squandered resources by bringing the provinces together to cooperate on a serious scale in the fight against fire was proposed. However, it was not taken up with eager enthusiasm by governments.

The N.R.C. study also brought together for the first time statistics on the public fire service in Canada as a whole, establishing that there were some 84,000 firefighters serving in 3,317 municipal fire departments. Career firefighters made up 14,000 of them. Although over 3,000 fire departments were wholly volunteer or paid a token stipend for their services, the largest part of the population and property value of the cities and towns of the country was protected by the career or partly-career forces. The largest fire department was that of Montreal, with a peak of over 2,400 firefighters and support staff in the latter 1970s, before a rationalization with attrition of manpower was begun to reduce the escalating cost of maintaining such a standing army.

The reduction of firefighting costs became a priority of municipal management, born of the economic difficulties of the late seventies and early eighties. This resulted in a close scrutiny of how the fire service operates and a demand for justification and measurement of methods that are based heavily upon firefighting. This is a labour-intensive job, in spite of the best of equipment, and very costly because of the necessity for having up to five firefighters on staff to keep one on duty at all times, 365 days per year.

Fire service leaders are aware that it would be better to prevent fires from happening than to wait for them to break out, then attempt to extinguish them. They are also aware of the fact that fires are occurring constantly and demand their immediate attention. To shift priority to the long-term reeducation of the Canadian public, making them as fire safety conscious as some other nationalities, and to overhaul our hazardous buildings will not be

Hamilton firefighters in modern protective gear training at their fire simulation centre. - D. Cossette, Hamilton Fire Dept.

The modern firefighter. — *D. Cossette, Hamilton Fire Dept.*

easy. There seem to be ingrained cultural factors behind our sorry international position. But such a shift seems essential on purely economic grounds due to the cost of strong firefighting forces.

In 1980 the fire chief heard himself referred to increasingly as a manager, less as the glamorous leader of his gallant forces going into the heat of battle. No longer was he the trumpet-wielding smoke-eater of old. The romance of firefighting was being succeeded by a demand for ability to handle budgets, computer programmes, and personnel questions tied to involved government regulations and union contracts. Yet the fire chief still has to make those fast, vital fire scene decisions. "Whichever way you decide, you will turn out to have been wrong," a veteran fire chief once said.

The advance of technology, generally, has produced many aids for the Canadian firefighter. At the same time it has confronted him with new and complex hazards. Where the citizen of the pioneer town stepped forward, without much know-how, to pit his musclepower against an enemy that threatened his neighbour's home or the whole community, many of today's firefighters are well-trained career professionals. At the same time there are still thousands of community-minded volunteers who leap from their warm beds to fight their neighbours' fires, and many of these are also well-trained and professional in their work.

Today, thanks to men of many generations who wanted to do a better job with better tools and in better ways, both the career and volunteer firefighters are able to bring a relatively sophisticated technology to bear on the old enemy. Above all, there is a strong tradition of humanitarian service, kept alive in a materialistic, cynical time by the real instinct that makes human beings rally to help one another in distress.

This characteristic has for many years spilled over into the field of cardiopulmonary resuscitation and other emergency medical responses. When full paramedical emergency services on a sophisticated level were developed in the United States and put into practice by the volunteer and career fire services, Canadian fire departments were ready too, but the necessary legislation was not forthcoming.

The fire service is now receiving more support from agencies, both government and private, that can help it in the increasing variety of emergencies. But it is still the fire service that unhesitatingly steps into the breach when some emergency arises that no one else can or will handle. Its non-fire-related emergency activities are still growing. When there is peril for life or property, fire, flood, earthquake or storm, the fire service has been, and probably will continue to be, in the words of an old fire company motto: "true to the call".

BIBLIOGRAPHICAL NOTES

Chapter II

Canada's first significant complex of cities and towns was the French colony of New France. There is probably no better general account of its background than Thomas B. Costain's *The White and the Gold, the French Regime in Canada*, particularly Chapters 9 to 13,covering its beginnings.

For a detailed history of Quebec City, the first city of New France, one work that regarded fires and key fire protection as matters worthy of mention is Pierre George Roy's *La Ville de Quebec sur la Regime Francaise*, Volume II. Roy quotes the *Gazette de Quebec* to document the first fire measure in Champlain's Abitation, the earliest regulations governing fire precautions, the Royal Well, in Upper Town, the proposed fire engine that never arrived, and the tough building regulations of Intendant Dupuy. Roy tells of the first major fire, in Lower Town, 1682, and describes the British bombardment and firing of the town.

The Lachine Massacre is described by Alfred Sandham in his *Ville Marie*.
The first firefighting organization, set up by Intendant Hocquart in 1734, is recorded in "Ordinances des Intendants," July 12, 1734, *Gazette de Quebec*, as quoted by Pierre George Roy.

Chapter III

The British era in Montreal is fairly well reported on the subject of fires and fire protection by Alfred Sandham in *Ville Marie, Sketches of Montreal, Past and Present.* Sandham's personal scrapbooks are in the Chateau de Ramezay Museum in Montreal.

E.Z. Massicotte, in *Bulletin des Recherches Historique*, Vol. XXX, covers some of the same ground, apparently through the same sources, in the Quebec *Gazette* for the "Ordinances of the Intendants," in putting forward his debatable assertion that the first Canadian fire department was formed in 1734, in Montreal, with Truteau as fire chief. He also uses the *Gazette* report on the meetings with Governors Beaucourt and Beauharnois to consider fire defense measures.

The acquisition of the fire engines for Montreal and Quebec in 1765 was announced in the Quebec *Gazette*, June 20. This was noted by Sandham and Vallee, along with the mention of Taylor as fire chief in Montreal.

Footprints of Assurance by Bulau, and Mainguy's history of the Quebec Fire Service, mention the involvement of the Quebec Fire Insurance Company in setting up a fire company with its name.

The Nova Scotia *Gazette* and Halifax Council minutes are from Public Archives of Nova Scotia (PANS), "Halifax Fire Companies," Vol. 209.

The earliest fire prevention measures in Halifax are recorded in papers collected in PANS Vol. 209, "The Halifax Fire Companies" and associated files, Halifax.

Quotations from the reports of Jenkin Jones, secretary of the Phoenix Fire Office in 1808, are used with kind permission of the Phoenix Assurance Group. These reports are a unique contribution to the urban Canadian fire protection picture we have of the day, though Jones could be very vague on the specifics of fire companies and engines.

The provision of a fire engine to Halifax by the Halifax Insurance Company is noted in the company's booklet *Halifax and the World.*

Saint John's first moves to provide regulation of fire hazards are recorded from early council minutes by Eric L. Teed in his *Canada's First City, Saint John.*

Reference to the Wellington Fire Club and the night watch is found in Veteran Fireman's account in the Colwell Collection, as are the descriptions of conditions in the department in the first half of the nineteenth century.

The ill-fated Guysborough settlement account is from George S. Brown's book *Yarmouth, N.S.*, written in 1818.

The Public Archives of New Brunswick (PANB) Doherty collection of material covers the beginnings of protection in Fredericton gleaned from a variety of sources.

The Miramichi Fire has been described in Brown's *Yarmouth, N.S.* and *Fire Waste in Canada* by Grove Smith.

Chapter IV

The burning of Newark, or Niagara, is told in *History of Niagara* by Janet Carnochan, perhaps more vividly than other accounts.

Among the few detailed records of individual fire departments is that found in John Ross Robertson's *Landmarks of Toronto*, Vol. II, Chapters 196 and 197. An explanation of the prominent newspaper publisher's special interest would seem to be evident in the appearance of his name among the members of boy firemen attached to one of the volunteer fire companies listed in Guillet's *Toronto, From Trading Post to Great City*. Guillet appears to make considerable use of Robertson's work in his account of Toronto fire history in the nineteenth century.

Edith Firth's *The Town of York, 1739-1815* adds to the earliest period, with the references to Alexander Wood, agent of the Phoenix, data from the Upper Canada *Gazette* dealing with the original fire engine in Toronto, and other references.

Chapter V

An authoritative source on North American fire marks is Bulau's *Footprints of Assurance*, based on the Home Insurance Collection in New York.

Several of the old British insurance companies, dating from the earliest days, have published their own histories; *The Guardian* and the *Sun* thus have described the organization of their own firefighting companies in London.

The Great Fire of London, as the event that triggered the establishment of organized firefighting companies and fire insurance, is described in a number of books, among which are *Heritage of Flame*, edited by D.J. Cannon, and *Fire Engines and Firefighting*, David Burgess Wise.

The dawn of post-Dark Ages fire engines is probably best covered by the 1850 text by Thomas Ewbank, *Hydraulics and Mechanics*, which seems to be the source of later treatments, including Cannon's excellent *Heritage of Flame*, and others.

Benjamin Franklin tells something of his efforts to begin fire insurance and fire protection measures in his autobiographical *Poor Richard's Almanack and Other Papers*.

The earliest American fire insurance companies and the era of the hand fire engine are particularly well documented in *Heritage of Flame*, beautifully illustrated and spanning this whole period, edited by D.J. Cannon.

Bulau's *Footprints of Assurance* mentions the Quebec Fire Insurance Company firemen and fire marks, as does Mainguy in the fire department's own record.

The role of the Phoenix Fire Office, evidently the first outside company to enter Canada, is presented in the Phoenix Assurance's booklets *First in the Field* and *Phoenix Family Story*, the material being used with their kind permission.

The detail on the British garrison role in firefighting in Halifax is found in the PANS Vol. 209, "The Halifax Fire Companies."

Chapter VI

The various newspaper articles and partial biography of Alfred Perry, active as a volunteer fireman, fire insurance inspector and city fire commissioner in the period of 1832 to 1900, as well as recollections of his nephew, William Perry, have provided considerable detail on the whole of the nineteenth century fire scene in Montreal. From this material, in the files of the late Montreal *Star* (notably the Carnival Edition of August 1884 and International Fire Engineers Edition of 1894), and in the collection of newspaper clippings and other material of A. Leslie Perry of Westmount, Quebec, much material in this and later chapters is drawn.

The descriptions of fires and fire engine contests from the diaries of Abraham Joseph are from a manuscript made available by kind courtesy of its editor, Annette Wolf, of Montreal.

The valuable contribution of the original handwritten reports by fire insurance surveyor John Josiah Broomfield on fire risks and fire protection conditions in a number of Canadian cities, including Montreal of the 1840s, were made available by the Phoenix Assurance Group, Toronto. This descendant of the first British fire insurance company also contributed their history in *First in the Field* and *The Phoenix Family Story*.

The LaPrairie fire story is from the Montreal *Pilot*, in a clipping with the Broomfield Reports.

Chapter VII

Stewart and Conwell, in their books on the fire of 1877, as noted for Chapter XII, describe the earlier fires in Saint John, as well. Some material on the fire of 1837 is found in the Leonard Scrapbooks in the Colwell Collection located in the New Brunswick Museum Archives, including the article "By an Old Fireman," from the Saint John *Globe*, December 14, 1901.

J.J. Broomfield's reports to the Phoenix also include descriptions and comment on Saint John, St. John's, Carbonear, Harbour Grace, Quebec and *Halifax.*

The development of the Halifax Fire companies in this period has been compiled from "The Halifax Fire Companies," Vol. 209, and articles from the Halifax Herald, Vol. RG1, Public Archives of Nova Scotia, as well as *Bicentennial of the Halifax Fire Department, a booklet prepared in 1968 by Pearl Connelly for the department, and Halifax and the World, 1809 and 1959*, a booklet published by the Halifax Insurance Company.

Chapter VIII

Copies of the original correspondence regarding donation of engines by the Phoenix to St. John's are included in the Broomfield report file of the Phoenix Assurance Group.

The mention of Jukes and minor parts of description of fires in this period are from R.G. Moyles' *Complaints is Many and Various But the Odd Divil Likes It,* as well as Broomfield's reports.

Chapter IX

Robertson quotes extensively from the minutes of the famous Toronto British America engine company, the engine itself still being in existence, apparently. The minutes cover the story of the Battle of the Don River Bridge. Although there has been confusion over the description of the engine in more recent accounts, Robertson correctly connects the "Montreal fore and aft" description to the Perry engine of the Provincial engine company.

For Hamilton, the Hamilton Fire Department's *A Century of Service, 1867-1967* is the major source, although Campbell's *A Mountain and a City* also supplied some of the material used here. J.J. Broomfield also had some things to say about Hamilton.

Ottawa's earliest history of firefighting organization is weak, but was researched and assembled in a special feature in the Ottawa *Evening Citizen* of August 30, 1949. Its sources were not indicated.

Many smaller cities and towns have no assembled history available, with limited glimpses only to be found in general histories of the community. Thus we find a little cameo of Cobourg in E.C. Guillet's *Cobourg, 1798-1948.*

Chapter X

Text concerning reformation of the old volunteer fire department and the best days of the Saint John volunteer companies is taken from material in the Colwell Collection, notably the article "By an Old Fireman" in the Saint John *Globe*, December 14, 1901.

Further evolution of the Ottawa, Halifax and Toronto fire departments is based upon the same sources as indicated for Chapters VII and IX.

Chapter XI

The story of the fire companies of Montreal in the heyday of the citizen volunteers, the burning of Parliament and the fire engine contests at the first two world's fairs are taken from the author's manuscript "The Fireman Who Burned Parliament," a biography of Alfred Perry. This is based on the various papers of Perry, with corroborative material from contemporary newspapers, such as the Montreal *Gazette*. The autobiography of William McRobie was also a good source on this volunteer period.

The great fires of 1850 and 1852 in Montreal are described in Alfred Perry's papers and Alfred Sandham's *Ville Marie, or Sketches of Montreal, Past and Present*. There are indications that Perry used Sandham to augment his memory of these events.

Chapter XII

Information on the new style of part-paid fire department in Saint John is provided by Conwell, the Colwell Collection and the Saint John *Globe*, December 14, 1901, the latter two in the New Brunswick Museum Archives.

When devastating conflagrations swept major cities like Boston, Chicago and Saint John in the latter nineteenth century, it was an occasion for quickly written books capitalizing on the widespread interest of the moment. In the case of Saint John, two such books recorded rich detail. These were *The Story of the Great Fire in St. John, N.B.* by George Stewart, a Saint Johner, as opposed to Col. R.H. Conwell of Boston, who wrote the competing *History of the Great Fire in St. John*. Both provide great detail on the buildings burned, with lists of the firms and households burned out.

The details of the progress of the Saint John fire and what each fireman did has been pieced together from the testimony given at the magistrate's inquiry following the catastrophe. This testimony is reported fully in clippings from the St. John *Daily Witness*, preserved in one of the Leonard Scrapbooks. E.P. Leonard was a steamer engineer in the Portland and Saint John

Fire Departments. The scrapbook is one of several in the collection of A. Montrose Colwell, a senior officer in the Saint John department in the 1940s. The collection is in the Archives of the New Brunswick Museum.

Material on Halifax for this time is also from the sources named for Chapter VI.

The changeover from the St. John's, Newfoundland, Volunteer Fire Department to a police-fire department has been described by former Provincial Fire Commissioner Frank Ryan in a paper published in *Firefighting in Canada*, December 1959. The conflagration of 1892 is also described by Ryan, as well as in several general history texts, including those of Prowse, Moyles and O'Neal.

Chapter XIII

The beginnings of the career fire service in Montreal are found in the descriptions by Alfred Perry and William Perry in the collection of A. Leslie Perry and in the City of Montreal Archives, annual reports of the Montreal Fire Department, 1863 to 1869, and *Fighting the Flames* by McRobie. These sources provide the principal data on the St. Urbain Street disaster.

The expansion phase of the Toronto Fire Department at this time, with the Osgoodby, *Globe* and Simpson's fires, is largely taken from the J.R. Robertson record in *Landmarks of Toronto*, Vol. II, Chapters 196 and 197 and the 1973 100th anniversary annual report of the fire department.

Events in late nineteenth century Ottawa are from the Ottawa *Evening Citizen* feature account of August 30, 1949.

Chapter XIV

Although the fire service history of the Prairies began long after that of the East, it is not particularly well detailed. Some of the major centres have a fairly well-detailed and lively story to be put together, but the smaller cities present little accessible history of interest.

A portrayal of the wilderness stage of the West, in the 1870s, has been vividly put forward firsthand by Reverend George Grant, secretary to Sir Sandford Fleming's C.P.R. route-surveying expedition across the continent, in his *Ocean to Ocean*.

The Hudson's Bay Company was the pioneer presence of civilization in the West in those times and the focus of the first communities. Purchase of the first fire engines for HBC posts is described in correspondence and the records of the company's proceedings, preserved in the Hudson's Bay Company Collection of the Public Archives of Manitoba.

The history of fires and the fire service of Winnipeg to the present is collected in the illustrated souvenir book *Alarm of Fire* by Vince Leah for the Firefighters' Burn Fund, Winnipeg.

Little early fire service history has been found on the cities and towns of Saskatchewan, and it is a short history relative to the East, certainly. The reluctance of the authorities to spend on fire protection in Regina is certainly clear, though, from various brief references.

The story of the Calgary Fire Department and early Calgary fires has been recorded in department files and the fine little book *Sirens and Cinders* by Carol Kobuta.

The Edmonton Fire Department's history has been compiled in the form of an archival file in the department, and some supplementary data is to be found at the Edmonton City Archives, notably in old souvenir booklets or convention programmes. *Edmonton, A History* by J.C. MacGregor has some useful references to the fire department.

Chapter XV

In *Ocean to Ocean*, George Grant, secretary to Sandford Fleming, describes Victoria and the settlements of the British Columbia Coast in the 1870s and earlier on a firsthand basis.

For most of the story of the volunteer era in Victoria, we are indebted to one of the few scholarly papers on the early fire service written in Canada, "The Fire Companies of Old Victoria" by F.W. Laing and W. Kaye Lamb, an unusually complete study.

New Westminster, the Royal City by Mather and McDonald and the account of the fire of 1898 in Grove Smith's *Fire Waste in Canada* were utilized in the New Westminster story.

Chapter XVI

The beginnings of Vancouver and its fire department in the nineteenth century comes from the fire department's various brief historical sketches in convention programmes, such as that of 1976, as compiled from old department records. The book *It Began with a Ronald* by Matches, and general histories like Eric Nicol's *Vancouver*, helped fill out the story.

The Dawson Museum kindly provided material from the Dawson *Daily News* and other material, including the small but well-detailed booklet *The Dawson Fire Department, Its Early History, Personnel and Efficiency*. *The Last Grand Adventure* by Bronson and Reinhardt provides a description of the conflagration of 1898.

Chapter XVII

The Ottawa - Hull conflagration of 1900 was the subject of a study by G.W. Shorter, Division of Building Research, National Research Council, with particular reference to the mechanisms of the spread of fire. The special feature on the history of fire in the Ottawa - Hull area, published in the *Evening Citizen*, August 30, 1949, provides considerable detail on this fire, apparently from the newspaper's files. Another description was given by Grove Smith in *Fire Waste in Canada*.

Some of the background on conditions, economic, physical and political, leading to this great conflagration is found in the interesting paper "The Conflagration and the City by John Weaver and Peter deLottinville, published in the *Urban History Review*. Yet another source of data on this fire is *The History of Insurance in Canada* by L.C. Vallee.

The Board of Trade fire in Montreal was also described by Grove Smith, but most of the present account is from an article by the author, researched from Montreal Fire Service files and contemporary newspapers. This was previously published in *Firefighting in Canada* magazine, June 1963, and is used with permission.

The Toronto Conflagration in 1904, again, was included in Grove Smith's fire accounts and was subject of a study by G.W. Shorter of the National Research Council. An interesting reconstruction of the fire's progress was presented by George Tatton in the *IAO Inspector*, published by Insurer's Advisory Organization.

The conflagrations in St. Hyacinthe and St. John's, Quebec, are described by Grove Smith and also by *Canadian Illustrated News*, September 16, 1876.

NFPA Quarterly, Vol. XII, and Grove Smith cover the Quaker Oats fire in Peterborough.

Chapter XVIII

The stories of Montreal in the early part of the twentieth century first appeared in *Firefight-*

ing in Canada magazine in June 1963 and are adapted for use here through courtesy of the editor, Don Glendinning. They are based on interviews with the late J. Gaston Tremblay, son of Fire Director Joseph Tremblay, the history files of the Montreal Fire Service, and the Archives, City of Montreal.

The history of Toronto in this era is derived from the Toronto Fire Department 100th anniversary booklet (1973), and the fine detail in J.R. Robertson's *Landmarks of Toronto*, Volume II, Chapters 196 and 197.

Calgary history covering the era of Chief Cappy Smart is from the files of the fire department and *Sirens and Cinders* by Carol Kabuta, written for the Century Calgary series.

The loss of the Parliament Building in 1916 is reported in *NFPA Quarterly*, Vol. 9 and Vol. 24, Grove Smith's *Fire Waste in Canada* (Appendix I), and various other sources.

The Halifax Explosion has been well documented, including *The Town That Died* by M.J. Bird; Graham Metson's *The Halifax Explosion*, with its collected reports; "The Harbour Catastrophe," a paper for the Dartmouth Heritage Museum; and the annual report of the Halifax Fire Department for 1910, courtesy of Audrey Manzer.

Data on the 1918 Montreal police and firefighter strike is from a 1919 report on the fire department by the Canadian Fire Underwriters Association, conversations with the late J.G. Tremblay, and the Archives, City of Montreal Fire Service files.

Chapter XIX

The great forest fires that occurred in the first part of the twentieth century are described by Grover Smith's *Fire Waste in Canada* and by Fred McClement in *The Flaming Forests*, the latter in considerable detail.

The Peshtigo Fire has been written about many times and is handled well by Paul Robert Lyons of NFPA in his "Fire in America."

Reports on the institutional fires in Quebec are found in the *NFPA Quarterly* as follows: Grey Nunnery - Vol. XX; Hospice St. Charles - Vol. XX; Laurier Palace Theatre - Vol. XXIV; St. Hyacinthe College - Vol. XXXI; Hospice St. Cunegonde - Vol. XI.

Chapter XX

The Commission on Conservation study of fire conditions in Canada, *Fire Waste in Canada*, written by J. Grove Smith, covers the establishment of provincial and federal fire commissioners and fire marshals, and the Dominion Fire Prevention Association. Others include: "Establishment of the Dominion Fire Prevention Association and Federal Government Fire Prevention Branch," a document of the Dominion Fire Commissioners's Office in 1954, and "60th Anniversary History of the Association of Canadian Fire Marshals and Fire Commissioners," prepared by that association in 1981.

Principal source for the following fires is *NFPA Quarterly*, as follows: Saint John Docks - Vol. XXV, as well as the Saint John *Evening Times Globe*, June 1931; the Cymbeline Disaster - Vol. XXVIII, also Montreal Fire Service files; Vancouver C.N.R. Pier - Vol. XXV; Pier D of CP Steamships - Vol. XXXII; Port Alfred pulp-wood fire - Vol. XXVI, and the report of Canadian Fire Underwriters Association, Montreal, May 14, 1932.

Chapter XXI

The war effort general account is based on descriptions by fire service personnel and the "Report of Managers on Work Performed for the Government of Canada as a Contribution to Canada's War Effort" by the Canadian Underwriters Association. The Montreal pier fire is reported in *NFPA Quarterly*, Vol. XLIV.

The episodes in The Fireboat War in Halifax are based on personal interviews with Commodore O.C.S. Robertson, R.C.N. (Ret.), and the late William Carson, former Fire Commissioner of Saskatchewan, as well as *Bicentennial of the Halifax Fire Department* by Pearl Connelly and *East Coast Port, Halifax at War* by Graham Metson.

Personal interviews with Frank J. Ryan, retired Fire Commissioner of Newfoundland, *NFPA Quarterly*, Vol. XXXVI, and the account in *Great Canadian Disasters* by Frank Rasky provided the bulk of the information for the story of the Knights of Columbus hostel fire in St. John's.

The Canadian Corps of Overseas Firefighters story is based on accounts by members of the corps, including now retired fire chiefs Joe Miller of Mississauga, Harold Doherty of Fredericton, and Denny Craig of Calgary. The operation is also mentioned in G.V. Blackstone's *A History of the British Fire Service*.

The Bedford Magazine incident was written principally on the basis of discussions with Commodore O.C.S. Robertson. It was also reported in the *NFPA Quarterly*, Vol. XXXIX.

Chapter XXII

The Rimouski and Cabano conflagrations are reported in *NFPA Quarterly*, Vol. XXXIX; the S.S. *Hamonic* in the same volume; S.S. *Quebec*, Vol. XLIV; and the S.S. *Noronic* fire in the official inquiry report by Justice R.L. Kellock for the Department of Transport.

The Time Building fire was reported by Western Canada Insurance Underwriters Association, August 1954, and *NFPA Quarterly*, Vol. XLIV.

The Hurricane Hazel account is from the Toronto Fire Department 100th Anniversary report in 1973, discussion with various fire department officers of Metropolitan Toronto, Rasky's *Great Canadian Disasters*, and other sources.

The author's own experience is the basis of the bulk of the material on development of building codes and fire codes in recent times, on fire service communications systems, introduction of the fog firefighting method, expansion of rural fire protection, and changes in fire apparatus in the post-World War II period. This is enhanced considerably by a variety of records and files of Fire Underwriters Survey and the Insurers Advisory organization.

Chapter XXIII

Statistics on volunteer and career fire departments are from the Switzer - Baird report *Study on Fire Prevention and Control Systems in Canada*. General conditions in the volunteer fire service are written on the basis of a long and inspiring personal relationship with this service.

The second St. Urbain Street Fire is based on a firsthand account by the author, previously published in part in *Firefighting in Canada* magazine, June 1963, and appears with the permission of the editors.

The fire in the Montreal Metro subway system was the subject of a detailed report by Canadian Underwriters Association, December 9, 1971.

The Mississauga train wreck and fire are described on the basis of a paper and articles by Fire Chief Gordon Bentley of Mississauga, and news media reports.

Fires in CIL House and the Chateau Champlain Hotel in Montreal are from reports by the author and published in *NFPA Quarterly, Vol. LVII, and NFPA Fire Journal*, Vol. 62, respectively.

The tragedy at the Bluebird Cafe Wagonwheel Lounge in Montreal is reported from the *NFPA Fire Journal*, Vol. 67, and that in the Opemiska Club, Chapais, from Vol. 71.

Firefighting during the strikes of 1971 and 1974, in Montreal, is described from personal exchanges with Eric Sprenger and others involved, as well as contemporary newspaper accounts.

Concluding statements on the state of public fire protection in Canada are those of the author, with some help from data in the National Research Council's *Study on Fire Protection and Control Systems in Canada*, 1980.

Sources

Arthur, Elizabeth, ed. *Thunder Bay District 1821-1892*. Toronto: Champlain Society, 1973.

Baird, Donal. "The Fireman Who Burned Parliament." Unpublished biography of Alfred Perry.

Baird, Donal. "Two Centuries of Fire Engines." Unpublished. Prepared for the National Museum of Science and Technology, Ottawa, 1970.

Baird, Donal. "A Century of Service." *Firefighting in Canada*, Vol. VII, No. 3, June 1973.

Bird, M.J. *The Town That Died*. New York: G.R. Putnam, Sons, 1962.

Blackstone, G.V. *A History of the British Fire Service*. London: Routledge and Kegen Paul, 1957.

Bronson, William and Richard Reinhardt. *The Last Grand Adventure*. Toronto: McGraw Hill Book Co., 1977.

Broomfield, John J. Reports and letters. See Phoenix Assurance Co.

Brown, George S. *Yarmouth, N.S.* Boston: Rand Avery Company, 1818.

Bulau, Alwin E. *Footprints of Assurance*. New York: The Macmillan Company, 1953.

Calbeck, Lorne C. *Charlottetown, Cradle of Confederation*. Fredericton: Brunswick Press, 1964.

Campbell, Marjorie Freeman. *A Mountain and a City*. Toronto: McClelland & Stewart, Ltd., 1966.

Canadian Fire Underwriters Association (now Insurers Advisory Organization). Various early special reports to members, Toronto.

Cannon, D.J., ed. *Heritage of Flame*. Garden City, N.Y.: Doubleday and Company, Inc., 1977.

Carnochan, Janet. *History of Niagara*. Toronto: Willliam Briggs, 1914.

Connelly, Pearl. *Bicentennial of the Halifax Fire Department, 1768-1968*. Souvenir booklet, Halifax Fire Department, 1968.

Colwell, A. Montrose. Collection of papers including scrapbooks of E.L. Leonard, Archives of New Brunswick Museum, Saint John.

Conwell, R.H. *History of the Great Fire of St. John*. Boston: B.B. Russell, 1877.

Costain, Thomas B. *The White and the Gold, The French Regime in Canada*. Garden City, N.Y.: Doubleday & Company, Inc., 1954.

Dartmouth Heritage Museum. "The Harbour Catastrophe." Pamphlet.

Dartmouth Union Protection Company. Centenary Booklet. Dartmouth, N.S., 1976.

Doherty, Harold F. Collection of papers. Public Archives, Fredericton, New Brunswick.

Dominion Fire Commissioner. *Establishment of the Dominion Fire Prevention Association and the Fire Prevention Branch*. Office of the D.F.C., Ottawa, 1954.

Dunshee, Kenneth H. *Engine! Engine!* New York: The Home Insurance Company, 1939.

Ewbank, Thomas. *Hydraulics and Mechanics*. New York: Bangs and Platt & Company, 1850. (Also a later edition under the title *Hydraulics and Other Machines, or Raising Water, Ancient and Modern*. New York: Scribner, Armstrong & Company, 1876.)

Firth, Edith, ed. *The Town of York, 1739-1815*, Vol. II. Toronto: Champlain Society, 1962.

Fitzpatrick, E.J. *The Dawson Fire Department, Its Early History, Personnel and Efficiency*. Dawson: G.M. Hill, 1902.

Franklin, Benjamin. *Poor Richard's Almanack and Other Papers*. Reading, Pa.: Spencer Press, 1936.

Glasscock, Nugent. *A Naval Sketchbook*. London: N. Glasscock, 1826.

Grant, Rev. George. *Ocean to Ocean*. Toronto: Belford Brothers, 1877.

Guillet, E.C. *Coburg, 1798-1948*. Coburg, Ont.: Business and Professional Women's Club of Coburg, 1948.

Guillet, E.C. *Toronto, From Trading Post to Great City*. Toronto: Ontario Publishing Company, 1934.

Halifax Insurance Company. *Halifax and the World, 1809-1959*. Halifax Insurance Company, 1959.

Hamilton Fire Department. *A Century of Service, 1867-1967*. Hamilton Fire Department, 1967.

Hanington, Felicity, with Captain Percy A. Kelly. *The Lady Boats*. Halifax: Canadian Marine Transportation Centre, Dalhousie University, 1980.

Jones, Jenkin. Reports and letters. See Phoenix Assurance.

Kellock, Justice R.L. *Report of Court of Investigation into the Loss of S.S. Noronic*. Ottawa: Department of Transport, 1949.

Kubota, Carol. *Sirens and Cinders, A History of the Calgary Fire Department*. Calgary: Century Calgary Publications. 1975.

Laing, F.W. and W. Kaye Lamb. "The Fire Companies of Old Victoria."*B.C. Historical Quarterly*, Vol. X, No. 1 (January 1946).

Lawson, J. Murray. *Yarmouth, Past and Present*. Yarmouth, N.S.: Yarmouth *Herald*, 1902.

Layman, Lloyd. *Attacking and Extinguishing Interior Fires*. Boston: National Fire Protection Association, 1952.

Leah, Vince. *Alarm of Fire*. Winnipeg: Firefighters' Burn Fund, 1982.

MacGregor, J.C. *Edmonton, A History*. Edmonton: Hurtig, 1967.

Mainguy, Cyrille. "Le Service de Protection Contre l'Incendie de la Ville de Quebec." Unpublished. Service de Prevention Incendies, Quebec, 1976.

Massicotte, E.Z. "Les Pompiers de Montreal sous le Regime Francais." *Le Bulletin des Recherches Historiques*, Vol. XXX (1924).

Matches, Alex. *It Began with a Ronald*. Vancouver: Mitchell Press, Ltd., 1974.

Mather, Barry and Margaret McDonald. *New Westminster, The Royal City*. New Westminster, B.C.: J.M. Dent and Sons and the Corporation of New Westminster, 1958.

McClement, Fred. *The Flaming Forests*. Toronto: McClelland & Stewart, Ltd., 1973.

McKinnon, J.C. *Old Sydney*. Belleville, Ontario: Mika Publishing, 1918.

McRobie, William O. *Fighting the Flames, or 27 Years in the Montreal Fire Brigade*. Montreal: Witness Printing House, 1881.

Metson, Graham. *East Coast Port, Halifax at War*. Toronto: McGraw Hill Ryerson, 1978.

Metson, Graham. *The Halifax Explosion, December 6, 1917*. Toronto: McGraw Hill Ryerson, Ltd., 1978.

Montreal City Archives. Dossiers on Montreal Fire Service and Annual Reports on Montreal Fire Department.

Morris, J.A. *Prescott 1810-1967*. Prescott, Ontario: Prescott *Journal*, 1967.

Moyles, R.G. *Complaints is Many and Various But the Odd Divil Likes It.* Toronto: Peter Martin Associates, 1975.

NFPA Quarterly. (Succeeded by the *Fire Journal.*) Boston: National Fire Protection Association.

Nichol, Eric. *Vancouver.* Toronto: Doubleday Canada, 1970.

O'Brien, Donald M. *A Centennial History.* New York: International Association of Fire Chiefs, 1973.

O'Neill, Paul. *The Story of St. John's, Newfoundland, The Oldest City, Vol. 2, A Seaport Legacy.* Erin, Ontario: Press Porcepic, 1976.

Perry, Alfred. Numerous newspaper articles and part of an unpublished autobiography in the possession of A. Leslie Perry, Westmount, Quebec.

Phoenix Assurance Company. *The Phoenix Family Story* and *First in the Field.* Company histories, undated. (Also the reports and letters of Jenkin Jones and John J. Bloomfield to the company, on file in Toronto.) Additional material on Jones and Bloomfield Reports are found in *Phoenix*, Vol. 1, A History of the Company published by Cambridge University Press, 1985.

Prowse, D.W. *Newfoundland.* London: McMillan & Company, 1895.

Rasky, Frank. *Great Canadian Disasters.* Toronto: Longmans Canada, 1961.

Robertson, John Ross. *Diary of Mrs. John Graves Simcoe.* Toronto: William Briggs, 1911.

Robertson, John Ross. *Landmarks of Toronto*, Vol. II. Toronto: Republished from the Toronto *Telegram* by J.R. Robertson, 1895.

Roper, Stephen. *Handbook of the Modern Steam Fire Engine.* Philadelphia: E. Claxton & Company, 1883.

Roy, Pierre George. *La Ville de Quebec sur la Regime Francaise*, Vol. II. Quebec: Queen's Printer, 1930.

Sandham, Alfred. *Ville Marie, or Sketches of Montreal, Past and Present.* Montreal: George Bishop and Company, 1870.

Shorter, G.W. *The Great Saint John Fire of June 20, 1877.* Fire Study No. 18. Ottawa: Division of Building Research, National Research Council, 1967.

Shorter, G.W. *Ottawa - Hull Fire of 1900.* Fire Study No. 7. Ottawa: Division of Building Research, National Research Council, 1952.

Shorter, G.W. *Toronto Conflagration of 1904.* Fire Study No. 8 Ottawa: Division of Building Research, National Research Council.

Smith J. Grove. *Fire Waste in Canada.* Ottawa: The Commission on Conservation, 1918.

Squires, W. Austin. *History of Fredericton — The Last 200 Years*, ed. J.K. Chapman. City of Fredericton, 1980.

Stewart, George. *The Story of the Great Fire of St. John, N.B.* Toronto: Belford Bros., 1877.

Switzer, R.A.W. and D.M. Baird. *Study on Fire Prevention and Control Systems in Canada. Ottawa: National Research Council, 1980.*

Tatton, R.J. "Toronto 1904 Fire." The IAO Inspector, No. 3, (1982). Insurers Advisory Organization.

Teed, E.L. *Canada's Oldest City, Saint John.* Saint John, N.B.: Lingley Printing Company, 1962.

Vallee, L.C. *History of Insurance in Canada.* Publisher Unknown, Library of Insurance Institute of Canada, Montreal.

Wolfe, Annette. "Abraham Joseph Diaries." Unpublished. Montreal.

Weaver, John and Peter deLottinville. "The Conflagration and the City." *Urban History Review*, Vol. VIII, No. 1 (June 1979).

Wise, David Burgess. *Fire Engines and Firefighting.* London: Octopus Books, Ltd., 1977.

Young, James A. *Reminiscences of the History of Early Galt.* Toronto: Hunter Rose and Company, 1853.

Photo and Art Credits

Photographs and drawings in this volume come from a wide diversity of sources. Apart from those supplied by the Public Archives of Canada and Public Archives of British Columbia, the great majority of pictures have been selected from the large collection assembled over many years by the author and directed towards this history project. The collection is drawn from the resources of numerous individuals, municipal fire departments, small museums and archives across Canada. All these sources have allowed copies to be made or supplied reproductions to assist in the author's objective of a series of books chronicling the history of the Canadian fire service.

Among the more prominent sources, organizations and individuals, are:

The fire departments of Victoria, Vancouver, Edmonton, Calgary, Regina, Winnipeg, Windsor, Hamilton, Toronto, Kingston, Ottawa, Montreal, Fredericton, Saint John, Dartmouth, Halifax, Yarmouth and St. John's.

The Public Archives of New Brunswick, Firefighters' Museum of Nova Scotia, Montreal Auxiliary Firemen, Winnipeg Firefighters Burn Fund, Dartmouth Heritage Museum, Chateau de Ramezay, Montreal City Archives, Phoenix-Continental Management Group, Halifax Insurance Company, Royal Insurance Group, Insurers Advisory Organization.

John Daggett, A. Leslie Perry, Eric Sprenger, Louis Stewart, Douglas Ferris, Paul Cleveland, Gaston Tremblay, Harold Doherty.

Particular appreciation is due for the fine original drawings done for the book by the Canadian artist and illustrator, Audrey Matheson.